D1221205

Dear Toney & Maggie
All our Love
Martin & Betty Ann
December 31, 2007

Waterworks

Waterworks

A Photographic Journey through New York's Hidden Water System

Stanley Greenberg

PRINCETON ARCHITECTURAL PRESS, NEW YORK

Published by
Princeton Architectural Press
37 East Seventh Street
New York, New York 10003

For a free catalog of books, call 1.800.722.6657.
Visit our web site at www.papress.com.

Printed and bound in China
06 05 04 03 5 4 3 2 1 First edition

Frontispiece: New Croton Dam, Westchester County, New York, 2000.

Typeset in Ionic and Twentieth Century.

Editing: Clare Jacobson
Design: Deb Wood
Layout: Nicola Bednarek

Special thanks to: Nettie Aljian, Ann Alter, Janet Behning, Megan Carey, Penny Chu, Russell Fernandez, Jan Haux, Mark Lamster, Nancy Eklund Later, Linda Lee, Nancy Levinson, Katharine Myers, Jane Sheinman, Scott Tennent, and Jennifer Thompson of Princeton Architectural Press —Kevin C. Lippert, publisher

Library of Congress Cataloging-in-Publication Data
Greenberg, Stanley, 1956–
Waterworks : a photographic journey through New York's hidden water system / Stanley Greenberg.
p. cm.
ISBN 1-56898-388-3 (alk. paper)
1. Water-supply—New York (State)—New York—Pictorial works. 2. Water-supply engineering—New York (State)—New York—Pictorial works.
I. Title.
TD225.N5 G74 2003
628.1'09747—dc21

2002015374

PREFACE

Stanley Greenberg

Waterworks play an important part in some of my favorite films. General Jack Ripper is obsessed with the fluoridation of water in *Dr. Strangelove*. Jake Gittes discovers how Los Angeles' watershed is stolen right out from under its owners in *Chinatown*. Waterworks are sites of drama in some other favorites. Harry Lime meets his end in the sewers of Vienna in *The Third Man*. Dr. Szell perishes in the Central Park Reservoir gatehouse (fictionalized for the film) in *Marathon Man*. And someone tries to murder Warren Beatty by opening up a dam spillway in *The Parallax View*. The impressive water infrastructure of these last three movies emphasizes the character's smallness or weakness or trivial role in the world.

The reality of New York City's water supply is as dramatic as these fictions. In fact, its size and scope would be unbelievable if not true. Along with eighteen reservoirs and three controlled lakes, there are also six balancing and distributing reservoirs, four tunnels to connect the reservoirs to each other, three aqueducts that bring the water to the city, and three tunnels that distribute the water around the city. The system contained other parts, now abandoned or given away. The Williamsbridge Reservoir in the Bronx, now a playground, was fed by the Byram Reservoir in Westchester, which still serves local residents. The Long Island system, which served Brooklyn and Queens, was partially given to Long Island towns. The rest of it was shut down because the wells it used were too polluted.

The city's main collecting systems are the Croton, Catskill, and Delaware. The three are interconnected—for example, the Delaware connects to the Croton under the West Branch Reservoir, while several "controlled" lakes feed into the Croton system of twelve reservoirs when necessary. Many people divide New York's waterworks into these three systems mentioned, but to the Department of Environmental Protection (DEP), the agency charged with supplying the city with water, it is just "East of Hudson" and "West of Hudson."

Even before I embarked on the photography for this book, I was fascinated by these systems and the web of interconnecting tunnels, gatehouses, reservoirs, and abandoned places covering (or covered by) several hundred square miles of land. I knew their basics from reading histories and guidebooks, and spending time in the library and in the DEP archives. I studied road and topographic maps for hints about where parts of the system were located. Soon I came to think of the system as an underground organism, like the giant fungus now regarded as the largest living thing on Earth.

Next I explored the systems themselves. I drove around upstate New York searching for signs of the operation. DEP employees who learned of my interest showed me unusual structures. I made photographs of parts that were open to the public. The first place I visited, in 1992, was the valve chamber at Shaft 2b in the new water tunnel. I had seen the chamber under construction several years earlier when I worked in city government. At the time it was just a huge space filled with construction equipment 250 feet under ground. When I visited again, the space was complete, but no water ran through it. The site was enough to make me want to see other parts of the tunnel, and gradually I became more interested in the system's history as well. Eventually I became able to "sense" the water system. Sometimes it was because of the way the road was paved, or the type of fencing along the roadway. I knew which buildings were part of the water system, whether or not they were marked. In 1998 the DEP allowed me to photograph the new water tunnel and some sites upstate. I was then closed out of the system, only to be let in again by the new commissioner. For reasons that I cannot guess, in 2000 I was finally given permission to photograph nearly every part of the system I wanted to see. While some of the structures were hidden deep in the forest, the watershed itself is not a secret. In fact, the city depends on people knowing where it is; many rules apply for activities in the watershed, and New Yorkers must depend on people following those rules if their drinking water is to remain unfiltered.

I completed work on this project early in 2001. After September 11, it became clear that neither I nor anyone else would be allowed the access that had been granted to me for a long time. When I looked to publish this collection, I encountered resistance from those who thought I was creating a handbook for terrorists. Fortunately, I found a publisher who believed, as I did, that it is important to publish these pictures. We have to be very careful about safeguarding our water supply, but we also have to stay connected to it. We need to protect it from terrorists, but we also need to protect it from farm runoff, automobile pollution, acid rain, household sewage, and many other forms of pollution that will only increase as development encroaches on the watershed. Only by knowing our system will we be able to insure the quality of our water.

The photographs in this book were made over a period of nine years. In that time, many people at the DEP helped me gain access to sites, gave me ideas for locations, and showed great interest in my project. Eileen Alter, John Bennett, Michael Cetera, Diana Chapin, Cathy Delli Carpini, Ted Dowey, Ben Esner, Amy Flavin, Marilyn Gelber, Mike Greenberg, Fred Grevin, Rafael Hurwitz, Al Miller, Natalie Millner, Joel Miele, Geoffrey Ryan, and Charles Sturcken all helped in different ways. Scott Foster was given the assignment of accompanying me on many of my visits; he opened many doors that I thought would not be available to me.

Many other people helped me along the way, in more ways than I can describe: Riva Blumenfeld, Richard Carboni, Judith Dollenmayer, Greg Conniff, Diane Cook, Bob Dawson, Catherine Edelman, David Eichenthal, Tom Garver, Rosalie Genevro, David Greenberg, Lenore Greenberg, Charles Griffin, Maria Morris Hambourg, Geoffrey Ithen, Len Jenschel, Barbara Kancelbaum, Jeffrey Kaye, Robert Klein, Gerard Koeppel, Ira Lyons, John Maggiotto, Ellen Manchester, Margaret Morton, Mario Muller, Candace Perich, Thomas Roma, Mitchell Schare, Bob Shamis, Rebecca Shanor, Bob Thall, George Thompson, Ken Weine, Alec Wilkinson, Sylvia Wolf, Sunny and Lloyd Yellen, and Jaye Zimet. Thanks to *Doubletake Magazine* for publishing some of the early pictures herein. My assistant Michael Altobello was always there to help figure out how to make a better picture. Yancey Richardson helped with her early support of my work and by providing a place to stay in the watershed. Of course, everyone at Princeton Architectural Press had a part in making the book, but I want to especially thank Kevin Lippert, Clare Jacobson, Deb Wood, Nicola Bednarek, Katharine Myers, and Nettie Aljian.

Finally, my wife, Lynn Yellen, and my daughter, Yelena, were and are the best friends and supporters any photographer could have. I look forward to making more books for them.

INTRODUCTION

Matthew Gandy

New York City is the most thirsty of all great cities.
—Jean Gottmann[1]

The provision of water for New York City is one of the most elaborate feats of civil engineering in the history of North American urbanization. As the city grew, it extended an "ecological frontier" of water technologies deep into upstate New York. The city's modern water supply system, which has been intermittently under construction since the 1830s, now extends across the largest water catchment area in the United States, collecting water from nearly two thousand square miles of sparsely populated mountains, lakes, and forests of the Catskill region along with the smaller and more densely populated Croton catchment in closer proximity to the city. The city's elaborate water infrastructure now includes nineteen collecting reservoirs, two city water tunnels, the world's largest storage tanks, and nearly six thousand miles of gravity-fed water mains. This vast network delivers 1.3 billion gallons of water a day to nine million people.

Some sense of the scale and complexity of the water supply system can be illustrated by the city's new six-billion-dollar water tunnel, City Water Tunnel No. 3, one of the biggest civil engineering projects in America, which was conceived in 1954 and has been under construction since 1970. Although the project was halted and nearly abandoned during the city's fiscal crisis of the mid-1970s, the first section was finally opened in 1998, and the whole tunnel is expected to be completed by 2020. Beneath an inconspicuous door in Van Cortlandt Park in the Bronx lies a 250-foot shaft linked to a dazzling subterranean valve chamber over 600 feet long that connects the city to its upstate sources of water supply. This extraordinary engineering achievement has an austere, utilitarian aesthetic reminiscent of the most impressive American water technologies of the twentieth century, such as Gordon B. Kauffmann's Hoover Dam and Eero Saarinen's Watersphere. At the tunnel's official opening in August 1998, a giant fountain was activated in the Central Park Reservoir that had first been used in 1917 to celebrate the arrival of water from the Catskill Mountains.[2]

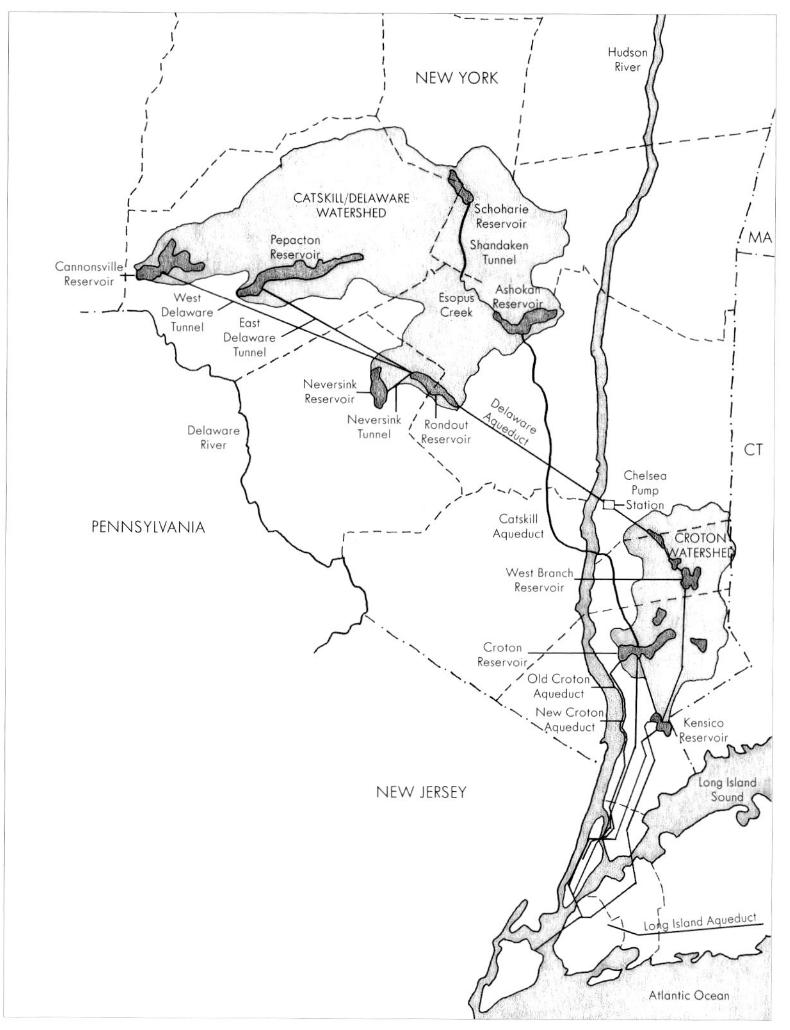

New York City Water Supply. *Jane Sheinman, based on map from NYC DEP.*

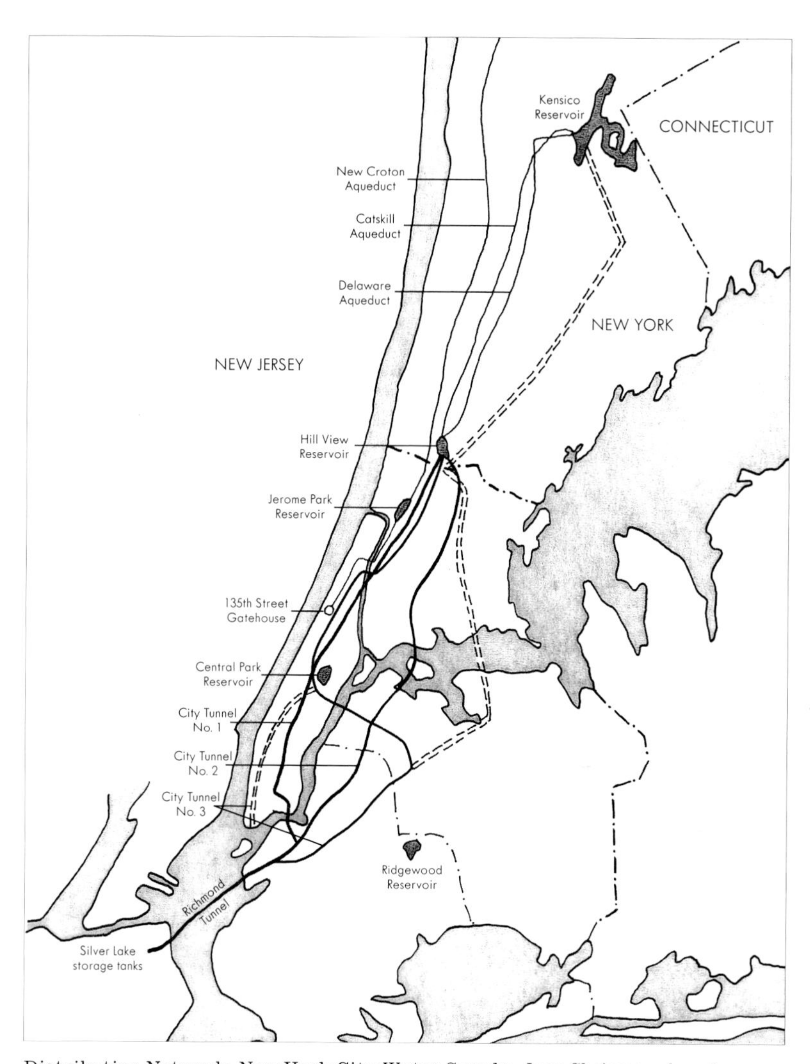

Distribution Network, New York City Water Supply. *Jane Sheinman, based on map from NYC DEP.*

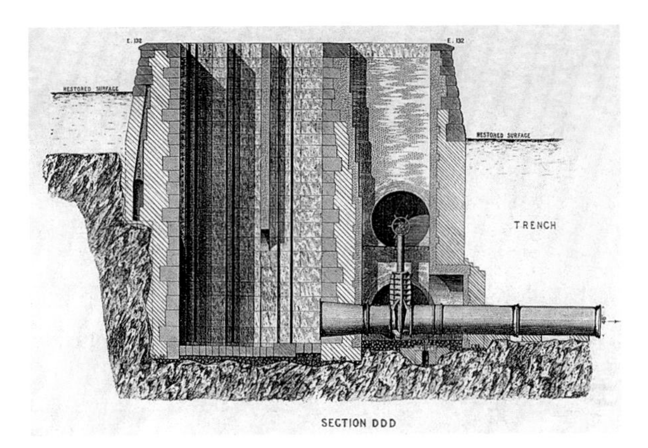

Construction drawing, gatehouse, New Croton Aqueduct, c. 1885, C. Gustafson, delineator. *From* Report to the Aqueduct Commission, 1887, *City of New York.*

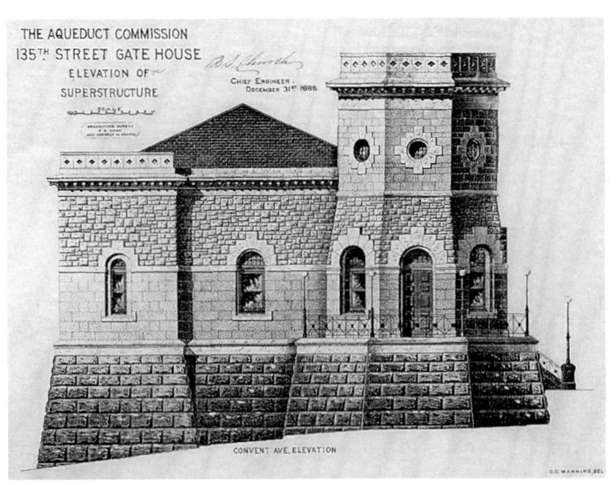

The Aqueduct Commission, 135th Street Gate House, Elevation of Superstructure, Convent Avenue Elevation, B. S. Church, chief engineer, C. C. Manning, delineator. *From* Report to the Aqueduct Commission, 1887, *City of New York.*

The history of cities can be read as a history of water. The historian Nelson Blake contends that "the indispensable precondition to the great growth of American cities during the nineteenth century was a recognition of the vital importance of water supply."[3] In the absence of plentiful supplies of water, cities are faced with the threats of fire, disease, social unrest, and material impoverishment. To trace the flow of water through cities is to illuminate the functioning of modern societies in all their complexity. Water is a multiple entity: it possesses its own biophysical laws and properties, but in its interaction with human societies it is simultaneously shaped by political, cultural, and scientific factors. For the historian Jean-Pierre Goubert, the modern era has seen a series of transformations in the use and meaning of water. The pre-modern waters of myth and salvation have been "subjugated, domesticated, mechanized and made profitable."[4] During the nineteenth century the water technology of the cities of Europe and North America evolved from an organic form, in which limited flows of water were combined with the harvesting of human wastes as fertilizers, into a modern hydrological structure in which far greater quantities of water were transported through the city in an ever more complex network of pipes and sewers. The control of water in the modern city now extends from the regional hydrological cycle to the application of plumbing technologies within the home. Water, in its multiple uses and transformations, flows between these different kinds of urban spaces linking diverse elements such as capital markets and domestic technologies into a multitiered social reality.

The story of urban water supply is a powerful element in the field of environmental history. Scholars have grappled with this dynamic from a variety of vantage points. In the American West, for example, the interaction between water and cities has become woven into a powerful narrative of the technological ingenuity behind urban growth. In New York, however, the relationship between nature and the urban landscape is far less clear. The architectural critic and landscape designer William Morrish, for example, suggests that the "dominant contextual elements of Los Angeles are not architectural, but natural." He describes how "the mountains surrounding Los Angeles are viewed as part of the formal vocabulary of the urban landscape," whereas New York, by contrast, is dominated by its architecture rather than its physical setting.[5] This text will show, however, that this comparison is misleading in ecological if not visual terms, since New York, like Los Angeles, is woven into its mountainous hinterland by an elaborate network of water technologies in order to transport billions of gallons of fresh water.

The landscape of upstate New York has been sculpted into a life-sustaining circulatory system through the interaction of the flow of water and the flow of money. Yet this double circulation of water and money is easily overlooked. The more distant parts of the city's watershed still resemble a Thoreauvian wilderness: one can trek through parts of the Catskill Mountains without encountering another human being. It is easy to imagine that you have entered a fragment of primal nature, but there are signs of human influence all around: the absence of large mammals; the patches of trees that have regenerated since the land was logged in the nineteenth century; and the network of paths that bear witness to many centuries of intense human activity. Most remarkable of all, however, is the fact that you are standing inside New York City's water system. The hydrological cycle of a whole region has been harnessed to provide water for a city: the rain dripping down through the leaves of hemlock trees will eventually find its way into the pipes and taps of millions of homes.

WATER AND THE NASCENT CIVIC REALM

Access to water was a constant problem for early settlers as they began to colonize the southern tip of Manhattan Island during the seventeenth century. In New Amsterdam, as the settlement was then known, there were frequent water shortages. Although many businesses and wealthier dwellings had their own wells, those without private water sources, principally the urban poor, depended on public wells, the first recorded of which was dug at what is now the intersection of Bowling Green and Broadway in 1658. The meager supplies from wells were supplemented by the collection of rainwater through the extensive use of cisterns made of masonry or wood.[6] With the growth of the city many of the wells became contaminated by privies, cesspools, and the drainage of dirty water from the streets. By the mid-eighteenth century the Swedish traveler and diarist Peter Kalm reported that even travelers' horses were reluctant to drink New York water. "There is no good water in the town itself," writes Kalm. "This want of good water is hard on strangers' horses that come to the place, for they do not like to drink the well water."[7] Records suggest that the only major source of good water remaining in the eighteenth century was derived from a spring outside the city near the Collect Pond at what is now Park Row. Water derived from a pump over the spring was sold to those who could afford it by water vendors known as "Tea-water Men." By 1785, however, the city's main source of water had degenerated into what the *New York Journal* described as "a very sink and common sewer."[8] In 1808 the city resolved to fill in the stinking Collect Pond in order to provide employment for some of the thousands of sailors and laborers thrown out of work by the Embargo Act. This early large-scale public works project reveals from the outset how political and economic factors would predominate in all discussions surrounding water supply and the construction of urban infrastructure. Rising land values merely accelerated the rate of development around the Collect Pond, and by 1815 the site had been completely filled in.[9]

Crowded, insanitary conditions in New York led to repeated epidemics of infectious disease. A series of outbreaks of yellow fever, a deadly virus carried by the mosquito *Aedes aegypti*, was recorded during the eighteenth century. Between 1791 and 1821 yellow fever epidemics became even more frequent. The most severe outbreak was recorded in 1798, when we find some of the earliest direct links made by city physicians between poor sanitation and disease. In the 1805 outbreak nearly half the city's population fled and the economic survival of the port of New York was threatened.[10] As for cholera, the great scourge of the nineteenth-century city, serious epidemics struck in 1832, 1849, 1854, 1866, and 1892 (the last major outbreak). The 1832 cholera outbreak, in which more than three thousand people died, was to prove pivotal in the early politicization of public health reform.[11] "On no former occasion," wrote the physician John W. Francis, "has New-York, frequently visited by the direful ravages of the yellow-fever, exhibited a more melancholy spectacle. Of a resident population of two hundred and twenty thousand . . . at least one-third are now dispersed in every direction."[12] Yet for Francis, as for most of his contemporaries, the cause of cholera remained inexplicable:

> It is conceded by all, that the origin of epidemic diseases is still enveloped in great obscurity; and the theories on this subject, whether referring to a distempered state of the atmosphere, to exhalations from putrid animal or vegetable matter, or to specific contagion, have been alike conjectural and unsatisfactory. The cholera, like all preceding epidemics, has exercised, but without any very useful results, the ingenuity of the speculative and philosophical observer.[13]

Francis described how cholera extended to "the innumerable circumstances connected with the economy of man in every state and condition" and attacked "rich and poor, native and stranger, young and old."[14] Despite the indiscriminate pattern of cholera morbidity across different social classes, the moralistic and superstitious responses to disease were only partially displaced by emerging concerns with urban sanitation. In the mid-nineteenth century we find a medley of perspectives on the cause of disease, combined to produce a moral geography of illness. In 1849, for example, the New York City Board of Health could confidently assert that "the general cause of the disease appears to exist in the atmosphere" and that "the agency of various exciting causes is generally necessary to develop the disease. Among these causes the principal are the existence of filth and imperfect ventilation, irregularities and imprudencies in the mode of living, and mental disturbance."[15]

In the prebacteriological era the problem of water supply was predominantly perceived to be one of bad taste and insufficient quantity; the danger and nuisance of poor water quality was not widely linked to disease. The cholera epidemics did, however, begin to widen and intensify public debate over urban sanitation, even if the precise mechanisms of contagion were imperfectly understood.[16] Commentators such as the physician Martyn Paine noted that the prevalence of cholera was worse in Paris and Montreal than in New York and began to develop a more rigorous analysis of variations in ventilation and cleanliness that might contribute to the severity of the disease.[17] The cholera epidemics of the nineteenth century were, above all, a transitional moment in the history of capitalist urbanization, as new trade routes exposed insanitary cities to the threat of disease before there were any concomitant advances in the science of epidemiology or the practice of public health. They were, in other words, the outcome of an uneven modernity that exposed a series of political, cultural, and scientific contradictions running through urban society. While it is undoubtedly the case that changing attitudes toward public health were a significant pretext for the modernization of the city's water supply (and environmental conditions more generally), these reforming impulses belonged to a wider political and economic context for capital investment in the physical infrastructure of cities. These changes rested on an emerging commonality of interests between the power of technical elites and the economic logic behind the reordering of urban space.

The late eighteenth and early nineteenth centuries were the worst period of all in the history of the city's water supply and heralded an increasingly panicky series of investigations into alternative sources. In 1774 the city's Common Council approved a plan put forward by the Irish-born engineer Christopher Colles to build the city's first municipal waterworks using a hilltop reservoir from which water could flow by gravity south to the town.[18] By approving Colles's scheme, the city embarked on a completely new approach involving the construction of a much more sophisticated water supply system than the existing wells and springs. Construction began in 1774 using a steam engine designed by Colles himself, connected to a network of water pipes made from pine wood. The city financed the construction by issuing five-percent bonds known as "water works money." In the event, the project was destroyed before any pipes were laid during the British occupation of the city in 1776. In spite of this setback, the city

Manhattan Company Reservoir, c. 1825. *From Edward Wegmann,* The Water Supply of the City of New York, 1658–1895 *(New York: John Wiley & Sons, 1896).*

was slowly but inexorably widening its responsibilities for water supply. In 1792, for example, it began to use tax revenues for the digging of new wells, so that by 1809 there was a network of 249 public wells.[19]

Before any collective solution could be found for the city's water supply problem, however, there was an extraordinary historical detour created by rival attempts to control New York's banking system. In the 1790s the only banks in New York were controlled by the Federalists under Alexander Hamilton. The leading Republican, Aaron Burr, knew that his attempts to set up a rival bank would be thwarted by his political opponents in the legislature. Burr successfully lobbied against the construction of a public water system on the grounds that the city would be unwilling or unable to adequately finance it. In 1799 he then succeeded in passing a bill that granted a charter to the Manhattan Company of which he was chair of the board. Hidden in the redrafted bill was a clause that allowed the company to use its surplus capital "in the purchase of public or other stock, or in any other monied transactions or operations."[20] Having set up the Manhattan Company, Burr rejected the more expensive options contained in the company's charter of diverting clean water from the Bronx River in Westchester County (which would have used up too much of the company's capital) and opted instead to dig a well close to the polluted Collect Pond and lay leaky wooden pipes. The company never constructed the steam pump and million-gallon reservoir that it had promised the city but relied instead on horse power in combination with a small reservoir one-tenth the size, adorned with a "false front of four Doric columns supporting a recumbent figure of Oceanus."[21] As well as being of poor quality, the water was also extremely expensive at twenty dollars a year, leaving most of the city reliant on rainwater collected in rooftop cisterns or "buying an occasional pailful from fetid wells."[22] Throughout its entire operations the Manhattan Company never laid more than twenty-five miles of water mains, but the profits were used to set up the Bank of Manhattan Company (later to become Chase Manhattan after the 1955 merger with Chase National Bank). So inadequate was the company's charter, which had been approved by the unwary State Legislature, that the city was forced to use its own revenues for flushing gutters, piping water to markets, and repairing streets after the company's laying of pipes. The city's attempt to buy the waterworks of the Manhattan Company in 1808 met with public indignation: not only had the company operated under the minimum legal obligations of its charter but it now stood to benefit financially from its own failure. In 1822 there was yet another serious outbreak of yellow fever, and fifteen prominent physicians signed a certificate warning that the Manhattan Company's water was unfit for human consumption. By 1830 the Manhattan Company still served no more than a third of the city: the rest of the population, who now numbered over 200,000, remained dependent on polluted wells or were forced to buy spring water from private vendors at exorbitant prices. The "Tea-water Men" and other water vendors were earning about $275,000 a year; ships were paying $50,000 a year to have their casks filled; and fire was destroying around a quarter of a million dollars' worth of property a year.[23]

Under the twin impetus of industrialization and immigration, New York was now the largest and fastest-growing city in America. A building boom in the 1820s and 1830s spread rapidly northward from the southern tip of Manhattan Island as former residential districts became converted into more lucrative commercial and industrial premises. Beyond the business district, with its dense concentration of new urban infrastructures such as sewers and gas lighting, the rest of the city languished in a state of anarchy. In the early 1830s public interventions multiplied with the construction of an iron reservoir on still-suburban Thirteenth Street and the laying of extra pipes in the driest parts of the city, but these efforts amounted to little more than tinkering at the edges of the problem.

During the early decades of the nineteenth century we can observe the emergence of a shift in the politics of water, which began to undermine the claims of private provision. This set in train a reformist urban vision based on a more sophisticated conception of the technical, administrative, and financial dimensions to public works. We should be careful, however, not to overstate the significance of public health concerns in relation to the broader political and economic dynamics behind the modernization of urban infrastructure. In 1798, for example, physician Joseph Browne had called for the creation of public waterworks from sources beyond the city's political boundaries, but the continuing intransigence of city authorities in the early nineteenth century stemmed from a reluctance to finance any program of public works that might significantly raise taxes.[24] In the meantime, however, other major American cities were busily abandoning their reliance on private water suppliers. As early as 1798, for example, Philadelphia had pioneered the development of public water supply, followed by Cincinnati in the 1820s.[25] Finally, in 1833, a Water Commission was appointed by New York State to undertake the first systematic study of the city's water needs, deploying the latest advances in the geological and engineering sciences. This more rigorous approach coincided with a major cholera outbreak and enabled a consensus to emerge among New York's political and business elite that a source of water from outside the city had to be found. A number of different proposals were debated at the time, including the damming of the Hudson River, the tapping of the Passaic River in New Jersey, and the construction of a canal from the Bronx River into Manhattan. After much deliberation, technical opinion agreed on a plan to divert water from the Croton River forty miles to the north of the city using a closed masonry aqueduct.[26]

How can we account for the shift in technical and political opinion that occurred in the 1830s? A first development was the increased seriousness of water-related disease outbreaks in New York at a time when new evidence from its economic rival Philadelphia revealed that the construction of an improved water supply system had successfully reduced the incidence of disease.[27] A further issue was the growing status and professionalization of engineers, who were able to convince political and economic elites that a more technically ambitious solution to the city's water problems had to be found. The city's reluctance to invest in the project on the grounds of cost was countered by new evidence collected by the state-appointed water commissioners that at least thirty thousand owners of building lots would be willing to subscribe to a new water service and easily cover its estimated construction cost of $5.4 million. Another critical factor to emerge in the 1830s was the growing extent of damage to property by fire, which threatened the city's burgeoning insurance industry. Fire damage was now costing the city more than any notional expenditure on a new water system, particularly in the wake of the "great fire" of 1835 in which 674 buildings were destroyed.[28] Since directors of leading insurance companies were represented on both the state-appointed Water Commission and the city's Common Council, this must have intensified the sense of urgency to take action. There was also at this time increasingly frantic lobbying by chemical works, breweries, tanneries, distilleries, hotels, sugar refineries, and other parts of the fast-growing New York economy which relied on pure and reliable water supplies and would be ruined if the city authorities failed to take action.[29] The situation was merely exacerbated by the gathering pace of industrialization, urbanization, and the use of steam power, which accelerated the shift of industrial production to new urban centers. In the absence of a secure water supply, New York's boosters claimed, the city would lose out to its main rivals and future investment would be stymied.

In 1834 the State Legislature finally passed a law that gave the city the right to construct its first municipally owned waterworks. This proposal was then strongly approved in a citywide referendum the following year. In 1835 the city authorized an initial bond issue of over $2 million which would allow construction of the Croton Aqueduct to proceed. The first bond issue sold well in both the United States and Europe, despite the economic downturn of 1837 to 1843, and the city was easily able to complete the project to the satisfaction of its investors.[30] Predictably, almost all the opposition to the new water system came from uptown residents and property owners whose wells were not yet polluted, and their well-organized opposition succeeded in delaying construction for a further two years. Toward the source of the Croton River in Westchester County, residents "vigorously opposed the aqueduct, claiming it disfigured their fields and divided property."[31] For the historian Eugene Moehring, however, the successful completion of this project demonstrated how "the city had triumphed over a water shortage that had threatened its health, property, and prosperity. Despite staggering costs and political opposition, authorities had met the crisis directly, setting an example for other towns. Problems had been legion—contract fraud, court suits, jurisdictional disputes, and pipe location, but New York overcame them."[32] The construction of the Croton Aqueduct marked a new era in North American urbanization. Had it not been built, it would have been impossible for New York City to retain its position as the largest and fastest-growing city in America.

ENGINEERING THE TECHNOLOGICAL SUBLIME

Nothing is talked of or thought of in New York but Croton water; fountains, aqueducts, hydrants, and hose attract our attention and impede our progress through the streets.... Water! Water! is the universal note which is sounded through every part of the city, and infuses joy and exultation into the masses, even though they are out of spirits.

—Philip Hone[33]

The completion of the Croton Aqueduct in 1842 was marked by the biggest public celebrations in New York City since American independence. The diarist Philip Hone recounts the five-mile-long procession as one of "perfect order and propriety," which he attributed to "the moral as well as the physical influence of water."[34] Since the writings of Vitruvius, the beauty and technical ingenuity of water-based architecture have been a recurring symbol of both prosperity and municipal independence. The architectural historian Vittorio Gregotti, for example, describes how the aqueduct has, through history, created "a productive dialectics with the built fabric of the city," which has enabled "the unity of the urbs and the civitas":

> To supply water freely to a city is much more than guaranteeing a service: it represents, in an exemplary way, the collective effort to ensure the communal life of the settlement. It imposes its necessary geometry and reconnects city and territory, geography and settlement. ...Necessity, ingenuity and civic virtue seem to be represented in the aqueduct by an organic synthesis.[35]

The creation of New York's water system consolidated the emergence of a more sophisticated kind of urban society within which fragmentary and parochial perspectives were superseded by a more strategic urban vision. This new outlook was reinforced by impressive engineering feats in the service of a modern metropolis. The Croton Aqueduct incorporated the latest advances in French and British engineering and also added unprecedented features of its own. The elaborate new structure excelled the Roman aqueducts in the size of its cross section and also advanced on Assyrian and Roman structures by the use of the siphon: its unique features included the low twelve-foot inverted siphons at the High Bridge and the one at the Manhattan Valley that put the water under pressure.[36] The chief engineer for the project from 1836 was John B. Jervis, who had gained extensive experience from the construction of the Erie Canal (1817–1825). Jervis was one of a number of transitional engineers who played a significant role in the modernization

Croton Dam. *From* Report to the Aqueduct Commission, 1887, *City of New York.*

Stereograph of High Bridge, c. 1900. *Collection of Stanley Greenberg.*

and professionalization of engineering science in the nineteenth century.[37]

Like the Erie Canal, the Croton Aqueduct also contributed toward a new kind of mediation between technology and nature in the American landscape. The economic growth of the city was increasingly tied to a regional urban ecology within which "wild nature" would be gradually displaced by an intensely reworked landscape. With the extension of the city's ecological frontier into upstate New York, the urbanization of nature and the naturalization of the urban could advance in tandem. The construction of aqueducts, dams, and reservoirs in upstate New York marked the evolution of a new kind of technological and cultural engagement with nature. "By the middle of the nineteenth century," argues the historian David Nye, "the American sublime was no longer a copy of European theory; it had begun to develop in ways appropriate to a democratic society in the throes of rapid industrialization and geographic expansion."[38] Crucial to this emerging aesthetic sensibility was the gradual supplanting of nature-based sublimity by an emphasis on the growing scale of human artifice. The new water infrastructures presented an architectural vision quite different from the kind of romanticized native American landscapes depicted in the contemporary art of Thomas Cole, Asher Brown Durand, and other artists associated with the Hudson River School of painting. The latest accomplishments of the engineering sciences emerged as a field in which America could rival the Old World in its Promethean transformation of nature. The vibrant combination of modernity and Arcadia presented a symbolic concretization of the pastoral landscape in the service of republican ideals. Features of the Croton Aqueduct such as the High Bridge soon became recognizable icons in the New York landscape and markers on a path to a more clearly defined sense of national identity.[39]

The construction of a new water infrastructure instituted a different kind of relationship between the city and nature. At one level the improved flow of water contributed toward a democratization of nature. Fountains and other architectural features symbolized the new urban bounty of fresh water brought from upstate sources; plentiful supplies of water helped to keep streets clean and contribute toward the creation of a more "hygienic" urbanism; while the gradual diffusion of plumbing technologies within the home lessened the daily burden of fetching water from standpipes and other sources. At first, some commentators were skeptical: the lawyer and diarist George Templeton Strong feared that Croton water would be full of "tadpoles and animalculae," to say nothing of

"Hibernian vagabonds" relieving themselves into the aqueduct as they toiled on the completion of the project. But within a year, Strong took delight in the new bathroom fixtures installed by his father: "I've led rather an amphibious life for the last week," he wrote, "paddling in the bathing tub every night and constantly making new discoveries in the art and mystery of ablution. A real luxury, that bathing apparatus is."[40] Water gradually entered urban consciousness in a variety of ways, some public and some private, and in time the growing use of water would be seen as an indicator of modernity.

Despite the better access to water for many ordinary New Yorkers, the political impetus behind the construction of the new water system remained firmly economic in its motivation. The historian Joanne Goldman, for example, suggests that the Croton Aqueduct marked a dramatic departure from the starkly unequal distribution of municipal services in the past, yet this observation oversimplifies the transformation of urban infrastructure.[41] The dramatic expansion in the scale and cost of public works in nineteenth-century New York was firmly grounded in an economic logic that found powerful political advocates. If impoverished Irish and German wards had received water before the wealthy residents of the Upper West Side, this was simply an anomalous outcome of the speed with which new pipes were constructed under the more densely populated parts of lower Manhattan.[42] The modernization of nineteenth-century cities in Europe and North America was not carried out in order to improve the conditions of the poor but to enhance the economic efficiency of urban space for capital investment. In this sense, the scale of new public works and the pace of technological change masked the persistence of social and political inequalities that would not be tackled in any systematic way until many decades later.[43] Advances in public health were an ambiguous by-product of the bourgeois rationalization of cities. Whatever the complex motivations that lay behind the development of elaborate public works projects, however, they did provide thousands of laboring jobs for native and immigrant workers whose votes were integral to the growth of machine politics with the widening of the political franchise. And even if the spread of new advances in public health was initially highly uneven, it did lend a powerful legitimacy to the development of new kinds of municipal governance freed from the noblesse oblige of the past.

So popular was Croton water that by 1850 New York could boast of the highest levels of per capita water consumption of any city in Europe or North America.[44] Earlier suspicions toward extensive water use were supplanted by a new enthusiasm for its diverse therapeutic and hygienic applications. Contemporary advocates of hydrotherapies such as Joel Shew saw the growing use of water as an indicator of the "general advancement of civilization" through American society.[45] Twentieth-century commentators such as Sigfried Giedion elaborated on this sense of dynamic optimism surrounding the spread of water through modern societies: "Words are too static," writes Giedion. "Only a moving picture could portray water's advance through the organism of the city, its leap to the higher levels, its distribution to the kitchen and ultimately to the bath."[46] Water use in American society grew steadily through the installation of new water technologies such as flush toilets and fixed washbasins, which came into general use from the 1850s onward. Yet the spread of plumbing technologies can be attributed to changing fashions in health and architectural design rather than simply the greater availability of running water. The new popularity of water within the home is best conceived as a "private manifestation" of the impetus for greater personal space and new standards of hygiene.[47] As the historian Alain Corbin has pointed out in a French context, the growing use of water and the concomitant emphasis on cleanliness formed part of a complex pattern of cultural changes in the pre-Pasteur era associated with sharpening social and economic differentiations.[48] Water use became entangled in wider ideological discourses surrounding the promulgation of middle-class domesticity in the face of increased social polarization in the nineteenth-century city.[49] After all, the "water revolution" was initially largely restricted to the middle classes, with most working-class tenements lacking bathing facilities until legislative changes in the early twentieth century. Only in 1870, for example, did the city finally open two free public baths after decades of debate over personal hygiene (private bathing houses had already been in operation since the eighteenth century).[50]

As for human wastes, there was little consensus over the relative advantages of water closets and earth closets until increased water use in the late nineteenth century began to overwhelm the use of facilities unconnected to the sewer system.[51] The declining use of earth closets in the cities of Europe and North America marked a transition away from the circulatory preoccupations of the organic city and the desire to use human wastes as fertilizers. As increasing quantities of water entered these self-contained sewage systems for individual dwellings, the nitrogen content began to fall. "It is now conceded," wrote the New York sanitation pioneer George Waring in 1895, "that the very small amount of manure and the very large amount of water cannot be separated at a profit."[52] With the advent of modern plumbing systems, these earlier efforts to recycle human wastes were superseded by a new emphasis on large-scale technological systems to facilitate the flow of water through cities, which led ultimately to the development of sewage treatment works and other advanced methods of water purification and pollution control.[53] The dwindling use of privies, cesspools, and other ad hoc solutions to the problems of urban sanitation marks the decline of the "private city" where emphasis was placed on minimalist and fragmentary forms of municipal government. The widening political franchise saw new legislative efforts in 1879 and 1901 to improve the quality of tenement housing and ensure that the benefits of new plumbing technologies would be incorporated into building design. New patterns of water use became part of a wider transformation in living conditions, which furthered the technical and cultural agenda of urban reformers and public health advocates as they sought to build centrally managed urban systems. For sanitary inspectors such as Robert Newman, the complexity of plumbing individual homes was a microcosm of the challenge to transform the circulatory dynamics of the entire city:

> No community and no city can preserve a wholesome condition without a supply of pure water; and an equally thorough purification from all refuse. To properly arrange this double circulation in a large house, is a matter of no trivial consideration; how much more, then, is skill, sagacity, and system, necessary for the sufficient supply and drainage of a district of an immense city like New York?[54]

By the early twentieth century the United States would be one of the best-plumbed nations in the world through the rapid diffusion of technologies such as pedestal sinks, enameled double-shell tubs, and siphonic-jet toilets.[55] An international survey of urban water consumption in the 1890s found that rates of water use in American cities were far higher than in those of continental Europe: only southern American cities such as New Orleans, with limited connections to urban water supply systems, exhibited lower rates of per capita water use than major European cities such as London, Paris, and Glasgow. We also find that New York's midcentury lead in per capita water consumption had been rapidly outstripped by fast-growing industrial cities such as Chicago, Pittsburgh, and Philadelphia.[56] Of course, the manufacturers of water-using technologies, plumbers, soap makers, and other industries that stood to benefit from an expanded use of water also played a role in fostering the spread of new consumption patterns: cleanliness was but one facet in the development of new cultures of retail shopping and advertising that would transform the lives of ordinary Americans.[57]

By the early 1870s a combination of droughts, low water pressure, and continuing urban growth, along with allegations of widespread graft and corruption, began to focus public attention on the need to rethink the scope and management of New York City's water supply system. As a result of the 1876 drought, for example, there was the first serious recognition that the repeated calls for water metering and repair of leaking pipes would not be sufficient to forestall serious future shortages. And by the 1880s supplies were so inadequate in the summer months that water could not be obtained from taps on the upper stories of tenement blocks. As a result of looming water shortages, a new phase of construction began for the Croton system in 1883. The New Croton Aqueduct (completed in

Building Extension of New Croton Dam, April 3, 1905. *From Report to the Aqueduct Commissioners 1895–1907, City of New York.*

New Croton Dam—Back of Croton-on-Hudson, New York, c. 1906, A. B. Kennedy, photographer. *Collection of Stanley Greenberg.*

1893), the New Croton Dam (completed in 1907), and the Croton Falls Reservoir (completed in 1911) were the most elaborate water infrastructures ever constructed up to that time. Had they not been undertaken, it might have proved catastrophic for the rapidly growing city: in 1895, for example, a severe drought had left most storage reservoirs almost empty.

By the early twentieth century the United States was the world leader in the building of dams for water supply, power, irrigation, and the control of navigation, yet as the historian Charles Weidner remarks, "most New Yorkers were too preoccupied with their city's growth to become unduly excited about their new aqueduct."[58] The New Croton Aqueduct was soon dismissed as simply one of the city's many achievements; it could hardly compare with the grandeur of other engineering projects such as John Augustus Roebling's design for the Brooklyn Bridge, which opened to international acclaim in 1883. The reconstruction of the "invisible city"—the upstate reservoirs, the underground pipe galleries, the valve chambers and other largely hidden or distant architectural features—could no longer capture the public imagination in the way they once had. "Municipal growth, whether slow or rapid," wrote the engineer James C. Bayles, "usually occurs by stages which are scarcely perceived, or at least scarcely realized, by citizens with whom it is a matter of daily experience."[59]

The completion of the Croton system did not solve the city's water problems. As early as the 1860s there had been fears for the safety of the city's new water system as a result of human and industrial wastes entering upstate streams and reservoirs. This marked an important advance in the epidemiological understanding of waterborne disease. In 1868, for example, Dr. Elisha Harris expressed concern over the "defilement" of the Croton system with sewage. Harris wrote widely on the subject and drew attention to the latest scientific debates in Europe and the need to disinfect drinking water by boiling.[60] A survey by the New York State Board of Health in 1884 revealed that villages, farmhouses, and mills in the Croton valley were draining their sewage directly into the river and its tributaries. As a result, the State Legislature granted the State Board of Health a series of new powers to control pollution in the city's watershed, but these proved relatively ineffective. Eventually, in 1893, new legislation was introduced that proposed to eliminate pollution by the acquisition of land along the streams of the Croton watershed. Despite contemporary descriptions of the water as turbid and strong-smelling, the city's newly established bacteriological laboratory assured the public in the early 1890s that the water presented no threat to public health.[61] But expert opinion remained divided between greater watershed protection and the early introduction of new filtration technologies. In 1894, for example, the engineer and entrepreneur Arnold Ruge lobbied the city authorities to introduce the latest French and Swiss filtration technologies on the pretext that the rich were simply buying themselves out of the problem of deteriorating water quality: "The rich in this city drink spring waters, imported from other States and even from Europe, but the masses of this City—the poorer classes—are compelled to drink unfiltered, dirty, and even odoros Croton water."[62] In the event, a wide range of changes were instituted, and the construction of the more distant and higher-quality Catskill system was begun, which allowed the city to avoid any need to filter its water supplies until the dramatic reemergence of the water quality debate in the 1990s. During the first two decades of the twentieth century the emphasis on improving or protecting water quality was advanced further with a variety of initiatives. The most significant change was the use of water chlorination from 1910, which led to a sharp fall in recorded cases of typhoid and restored public confidence in the safety of the city's water.[63]

With the rising status of engineering, planning, and other new public service professions, urban management became increasingly influenced by a technical elite devoted to the rationalization of cities. In the wake of the Chicago exposition of 1893 and new developments in city planning pioneered in Sweden, Germany, and other industrialized countries, there was a strong convergence of international technical opinion around the need for more sophisticated and scientifically based modes of urban governance.[64] By the early decades of the twentieth century engineers like George T. Hammond were celebrating their role in the "wonderful growth of the urban." For Hammond, writing in 1916, there was no doubt that engineers must play a didactic role: "We are employed by the public, not only to do their work but also to lead them in technical municipal affairs. It is our duty and province to instruct as well as serve."[65] The rationalization of urban space became a kind of Taylorization transferred from the factory to the city: the creation of a scientifically organized world where Thorstein Veblen's dream of "engineers in power" might ultimately be realized.[66] It would be misleading, however, to conceive of the growing professionalization of technical opinion as constituting an undifferentiated perspective on the future of city management: important differences existed over rival technical and organizational solutions to the management of urban space.[67] Debates over technology and urban planning are best perceived as crisscrossing a complex web of evolving interests in the context of continuing elite domination of urban politics. Tensions existed between rival fractions of capital and also between different tiers of state authority as successive political machines vied for control over urban government.[68] The 1898 merger between the five boroughs expanded the scope and responsibilities of New York City's water supply system overnight, as the Croton system began to replace inferior alternatives such as the polluted wells of Queens, Staten Island, and the Bronx, as well as Brooklyn's surface-fed Ridgewood system based in Long Island.[69] From the 1890s onward a policy consensus gradually emerged that the Croton system could not sustain any major expansion in the future and that the only solution to the city's long-term water needs would be to use the more distant sources of the Catskill Mountains.

Urban growth instituted a brutal logic of its own, which necessitated a transformation of the physical landscape over a vast area, introducing forms of strategic decision making beyond the scope of existing municipal government. An emerging political ecology of power linked the city to an ever greater swath of upstate land as part of a giant metabolic urban system. In order to begin this new phase of construction, the State Legislature created a powerful new structure in 1905 called the Board of Water Supply, which institutionalized the role of engineers in municipal government. This new body had immense powers: it could take over private land for water supply; it could work in relative autonomy from elected municipal government; and, following amendments to the state constitution, its projects were not limited in cost by the requirement to restrict bond issues to a specified percentage value of city real estate.[70] Foreign observers such as the British engineer Gilbert J. Fowler were amazed at the scale of this new undertaking:

> The world has been startled by the magnitude of your water schemes. Any European city would have regarded 114 gallons per capita as an extravagant allowance, yet this—which is equal to a daily supply of 500,000,000 gallons—is what is now obtained from the Croton works alone, but rather than curtail that supply, or do anything which might be interpreted to favour a limited use of water for public health purposes, the authorities determined to carry out a gigantic scheme to obtain another 500,000,000 gallons of water a day, this time from the Catskill mountains. The magnitude of the undertaking and the aggregate cost of bringing in 1,000,000,000 gallons per day will place New York water supply in a category by itself when a history of the world's great water works comes to be written.[71]

Before New York could commence an expansion of its water supply, however, it had to overcome what Charles Weidner describes as a "scandalous prelude" created by the attempt of the Ramapo Water Company to gain control over the water resources of the Catskill Mountains in anticipation of the city's future needs.[72] In 1895 the Ramapo Water Company had successfully lobbied the State Legislature to grant it land and water rights in the Catskill Mountains. As a result, New York City came very close to losing any control over the future design or cost of its water supply system. When the nature of the Ramapo Water Company's proposals became known, a bitter public response ensued led by City Comptroller Bird S. Coler.[73] In 1901 the State Legislature repealed the company's charter, thereby freeing the city from dependence on it for the future of its water supply. For some years, however, the company continued its attempts to overturn the decision of the State Legislature. As late as 1915, for example, the US Supreme Court dismissed a suit brought by the Ramapo Water Company that sought to prevent the construction of the Ashokan

Aerating the Water, Showing Gate Chamber, Ashokan Reservoir, c. 1920, J. Rubern, Publisher, Newburgh, N.Y. *Collection of Stanley Greenberg.*

Reservoir, the first of the city's reservoirs to be constructed in the Catskill Mountains.

The first part of the Catskill system, constructed between 1907 and 1917, was far bigger than the Croton system. It includes the Ashokan, Kensico, Hill View, and Silver Lake reservoirs, as well as 126 miles of aqueduct (some 18 miles of which were bored through solid rock between 200 and 750 feet under the Harlem and East rivers and the streets of Manhattan).[74] Under the McClellan Act the city sought to acquire land without a repetition of the "rapacious proceedings" that characterized the use of land in the Croton watershed, yet bitter conflict ensued over the indirect loss of earnings from destroyed businesses and undervalued property acquisitions. The development of New York's water system led to the mass displacement and destruction of many settlements across the city's watershed. Upstate New York experienced its own water wars to rival that of the Hetch-Hetchy and Owens valleys in California.[75] In 1908, for example, Justice A. T. Clearwater publicly reprimanded the city for the betrayal of the people of Ulster County, whose land claims had been "scoffed and sneered at, derided and belittled." In the summer of 1913 a correspondent for the Kingston Freeman described the

Rondout Pressure Tunnel, 14 feet, 16 inches, in finished interior diameter, during construction, with a portion of concrete lining in place. *Work of the Board of Water Supply: A General Description of the Catskill Water Supply and of the Project for an Additional Supply from the Delaware River Watershed and the Rondout Creek. Board of Water Supply of the City of New York, 1940.*

disappearance of the village of West Shokan to make way for the Ashokan Reservoir: "Very few buildings are left now to be burned. The trees are all cut down and the village is fading as a dream."[76]

The Ashokan Reservoir alone covers an area of 12.8 square miles, equal to the whole of Manhattan Island below 110th Street, with a capacity of over 130 billion gallons drawn from a mountainous catchment area of 257 square miles. The Catskill Aqueduct is twice as long as the greatest Roman aqueduct, being over twice the length of the two Croton aqueducts combined, and was designed by "a corps of engineers and experts unequalled in the history of engineering."[77] In 1917 Mayor George McClellan compared the construction of the Catskill Aqueduct to that of the Panama Canal. The comparison is apposite, considering the fact that the possibilities for large-scale urban reconstruction were facilitated by growing US economic and political hegemony in central and southern America at the time. Urban infrastructure provided a reliable investment for new flows of capital, and continued urban growth was in turn related to vast social and environmental transformations that extended far beyond the US frontier.[78] In October 1917 the first Catskill water reached Manhattan and this event, like the completion of the Croton Aqueduct in 1842, was marked by a three-day celebration culminating in a rapturous reception in Central Park:

> At the Sheep Fold ceremony depicting American Indian tributes to "the good gift of water" a huge chorus of children and young women sang the National Anthem and the pageant was about two-thirds through when a downpour came. Strangely the rain fell just after the medicine men of the Indian Village and the priests of the ancient Orient, in compliance with the programme, had prayed for rain for benefit of the crops. The prayers had no sooner been uttered than the drops of water heralded the cloudburst. The children, laughing at the coincidence, scattered in all parts of the park.[79]

Almost as soon as the Catskill system had been completed in 1927, however, it became clear that an even bigger water source would be needed for the city. During the 1920s the Board of Water Supply carried out "endless investigations" of the more distant parts of the Delaware watershed.[80] In 1930, in anticipation of the city's water needs, the state of New Jersey sought an injunction in the US Supreme Court in order to prevent New York City from taking further water from the tributaries of the Delaware River (whose water supplied the needs of many New Jersey communities). In the event, the court found in the city's favor, with a decree granting the city permission to take 440 million gallons a day from the Delaware River as it passed through New York State.[81] In 1927 the city's Board of Estimate finally approved the Delaware project and authorized the issue of $64 million worth of city bonds, but the 1929 stock market crash delayed the start of construction until 1937.[82] The first phase of the Delaware system was eventually completed with a federal loan issued under the public works program of the New Deal, but the entire project was not completed until 1967. It is difficult to overestimate the significance of the New Deal for the modernization of the city's infrastructure: in 1940, for example, the chairman of the New York City Planning Commission, Rexford Tugwell, estimated that New Deal–funded capital improvements amounted to double what could have been achieved without federal assistance.[83] Thus the initiation of the final stage of the city's water system was enmeshed in a wider political and economic context for public works, which increased the power and scope of large-scale infrastructure projects in urban policy making.

The period from the 1880s until the 1960s saw a continuous program of dam and reservoir construction for New York City. The idea that anything might restrict or impede the preeminence of New York as a world-class city proved unthinkable, and engineers were ready to provide ever more elaborate means to slake the "thirsty metropolis."[84] It would be misleading, however, to argue that engineers and city planners uniformly supported the logic of ever greater water use: there was a parallel ascetic discourse of disdain for the profligate use of water in the post–World War II era. The city engineer Edward J. Clark, for example, publicly derided the wastage of water by lawn sprinklers and children playing with fire hydrants.[85] The serious drought of 1949–1950 not only reinforced the urgency of the need to complete the Delaware project but also heralded intense water conservation efforts for the first time. By the early 1960s a further series of droughts necessitated the pumping of extra water from the Hudson River, and serious debate ensued over the mooted construction of nuclear-powered desalination plants as the only viable long-term water strategy for the city.[86] In the summer of 1965 the engineer Abel Wolman made a pointed contrast between the apparently drought-induced plight of New York and the increasingly sophisticated water infrastructures of semiarid southern California. Wolman railed against "delayed action and failures of management," which had led to the absurdity that while "New Yorkers were watching their emptying reservoirs and hoping for rain, Californians were busy building an aqueduct that would carry water some 440 miles."[87] For Wolman, water shortages in the modern era were the outcome of political vacillation rather than climatic perturbation.

HYDROLOGICAL TRANSFORMATIONS

The water supply of New York has passed through a series of transformations since the early seventeenth century. The first period, lasting from the founding of the original settlement in 1626 until 1658, was marked by a reliance on natural water sources and private wells. A second phase, from 1658 until 1774, saw an expanding network of public wells within a context of steadily declining water quality. A third interval, from 1774 to 1830, was dominated by a series of ill-fated private interventions including the role of the infamous Manhattan Company. This chaotic urban scene was characterized by repeated outbreaks of disease, uncontrollable fires, and escalating economic disruption. A fourth phase, from 1837 to 1911, saw the construction and expansion of the Croton system as the city's first comprehensive public water supply. The modernization of the city's water system was marked by a series of advances spanning the bacteriological, technical, and administrative dimensions to water resources management, which mirrored developments elsewhere in Europe and North America at this time. A fifth period, between 1907 and 1967, marked the completion of the Catskill-Delaware system and an expanded role for municipal government in the management of regional water resources. This was the zenith of the technical management of urban space, with maximum power and autonomy for government agencies reached under the New Deal era. The most recent phase, extending from the late 1960s until the present time, has been characterized by a series of complex challenges to existing patterns of water provision. Regional economic change and new patterns of sociospatial restructuring have contributed to the emergence of a series of major policy dilemmas in the fields of capital investment and water quality.

The period between the completion of the Croton Aqueduct in 1842 and the city's fiscal crisis of 1975 marks a phase of remarkable stability in the history of New York's water supply. The nineteenth century saw a decisive shift from private to public water provision in order to allow new levels of efficiency and coordination. A series of tensions were played out not only between public and private interests but also among disparate bodies of technical expertise and rival political machines. During the twentieth century some of these disputes were resolved with a move toward the greater consolidation of fragmentary interventions to form powerful regional systems of management and control. This implied a partial waning of local democratic input, as technical elites emerged to design and operate vast public works systems. The creation of semiautonomous government structures fiscally and politically insulated from local electorates marks a smaller-scale precursor to the powerful regionally based federal agencies of the New Deal such as the Tennessee Valley Authority. The dams, reservoirs, and other large-scale infrastructure projects of the New Deal era have been widely interpreted as the epitome of American modernism. The combination of a utilitarian aesthetic in the International Style with a functional commitment to the rationalization of regional water resources became a symbol of a new kind of public landscape. These "democratic pyramids," to use Lewis Mumford's phrase, represent a unique conjunction of technology, nature, and public policy making, but their physical longevity belies the fragility of the cultural and political circumstances that facilitated their construction.[88] With the fading of the New Deal ethos in the 1970s, a new set of political, economic, and cultural developments began to shape the evolution of regional water policy. Recent changes are distinctive in a number of respects: the emergence of new sources of environmental risk such as cryptosporidiosis; the development of greater degrees of public skepticism toward technical and scientific expertise; the weakening of city power in relation to regional political developments; and above all, the intensity of the neoliberal challenge to the fiscal autonomy and ideological legitimacy of an effectively regulated and adequately funded public water system.

For over 140 years New York City successfully provided cheap, plentiful, and high-quality water to its citizens, on the basis of a settled relationship between water technologies and the "democratic urban landscape"; yet this historic achievement is now thrown into doubt by a series of political and economic developments beyond the reach of any regulatory or democratic structures yet devised. Some urban scholars have argued that social and economic developments since the 1970s have lessened any technical link between capital and urban form: the connection between urban morphology and economic function has become weakened.[89] In the case of water supply, however, this claim is problematic because of the continuing functional dimensions to urban space. The ongoing construction of the city's third water tunnel, for example, suggests that the material determinants of urban form may conform to a deeper logic than the more ephemeral political and cultural shifts surrounding the construction and design of real estate and other speculative elements in the built environment. Still, even if the technical dimensions to the design of urban space retain a high degree of continuity, the pressures to transfer public assets into the private sector have become immense.

During the 1990s the privatization of urban water systems gathered global momentum. In 1997, for example, the *Financial Times* proclaimed, "Water is the last frontier in privatisation around the world."[90] The sale of public water systems not only flows from the fiscal weakness of municipal authorities worldwide but has also been pushed by national governments in order to bolster foreign currency reserves and find favor with international financial institutions. The global marketization of water has not been without high-profile protests, as grassroots campaigns in Argentina (Tucumán province), Bolivia (La Paz and El Alto), Manila, and Barcelona attest.[91] While the New York case did not lead to mass protests, there is little doubt that the city's water supply has been politicized to a greater degree than at any time since the failure of the Manhattan Company in the early nineteenth century. What we are seeing in New York is a protracted process of reshaping the role of the municipal government in urban water supply. In effect, a hollowing out of government arising from a combination of fiscal and ideological pressures is leading to a polarization in the public policy debate between demands for water quality protection, advanced principally by urban environmentalists, and a coalition of antiregulation upstate interests, whose rhetoric is rooted in a legacy of land use conflict in the city's watershed.[92] The future form of environmental management is emerging as a politically contested reconfiguration of public policy, in this instance centered on a redefinition of the administrative powers of city government. At the heart of the debate over environmental management in the city lies a tension between market-led development pressures and the administrative jurisdiction of municipal authorities. The blocked sale of the city's water system in 1997 suggests that a stable new configuration of power between capital and municipal governance has yet to be determined. Beyond the international political and economic exigencies that have driven recent developments in water policy, there is still considerable scope for contestation and debate. If filtration of the city's entire water system does eventually occur at some point in the twenty-first century, historians of the future may well comment on the remarkable persistence of this particular fragment of engineered nature.

The extensive dam- and reservoir-building program undertaken by New York City caused wide-ranging disruption to the communities of the Croton and Catskill watersheds, yet the extending ecological frontier of the city enabled a new kind of mediation between nature and society that was of inestimable benefit to millions of people. Municipal-led policy interventions under the auspices of technological modernism have often had deleterious environmental consequences, as the highway-spliced inner-city neighborhoods of postwar New York attest. Yet to dispense with the role of government altogether as part of an ecological critique of the institutional basis to Western modernity risks the effective abandonment of any practical means for implementing environmental regulation. This political and ecological dilemma is heightened by the global dimensions to environmental change, which are driven by the relative absence of any form of effective international economic regulation in the face of unprecedented capital mobility. The fact that global climate change may affect the hydrological conditions for New York's water supply in the future illustrates this profound uncertainty.[93]

Some recent critical interventions under the auspices of the postmodernity debate have tended to denigrate or at least display profound ambivalence toward the regulatory role of government in modern societies. The demise of the nation-state is both predicted and welcomed as part of a new fluidity in cultural and economic life. "On the ethical front," writes the anthropologist Arjun Appadurai, "I am increasingly inclined to see most modern governmental apparatuses as inclined to self-perpetuation, bloat, violence, and corruption."[94] This kind of antistatist or even conspiratorial sentiment is a recurring motif in environmental histories that are critical of urban demands on rural water resources. In an American context, for example, the water wars of the Midwest and southern California have proved a fertile ground for what we might term antimodern interpretations of large-scale state-directed water projects.[95] What is often missing from these accounts, however, is a fuller picture of the impact of the modernization of water infrastructures on the everyday lives of urban citizens. The provision of water remains a collective service, even if the public-private distinction has become sharper in recent years and even if the very word "public" misleadingly elides dominant economic, political, and cultural developments to the exclusion of more marginal voices. Water is a collectivity in a metabolic sense because urban life depends on its supply, but decisions over water policy have never been open to much in the way of public deliberation or debate. The most promising solution to environmental degradation may lie in the development of a more sophisticated public sphere through which new forms of democratic decision making can emerge in preference to any lurch toward the ecological Hobbesianism of greater control, which may prove in any case to be fiscally and ideologically untenable.

The role of strong advocacy groups and an informed and active citizenry emerge as crucial in any effort to protect the environmental advances of the past. Yet the post–New Deal environmentalist agenda harbors innate weaknesses: its individualist and consumer rights–based orientation serves to deflect attention from more widely conceived regulatory goals in the public interest that extend to the sphere of production as well as consumption. Similarly, the degree of indifference on the part of city-based water quality advocacy groups to the economic viability of low-income upstate rural communities is testament to wider class-based tensions in the American environmental movement, which serve to strengthen the hand of capital in the dismantling of the public sphere. This political dilemma is heightened by the socially regressive consequences of market-led ecological modernization, epitomized by spiraling water charges, that threaten to fragment the political strength of any cross-class environmental alliances in the city.

This text has explored the evolving interaction between water and the dynamics of capitalist urbanization. We have seen how the creation of urban infrastructure has been essential to the economic viability of New York City and at the same time has fostered possibilities for new kinds of mediations between nature and society. The period from the 1840s until the 1970s marked a longue durée in the history of the city's water marked by a high degree of political and organizational continuity in spite of rapid urban growth and far-reaching technological change. The partial unraveling of existing relationships between water and urban form since the 1970s reveals the fragile and contradictory dimensions to the built environment within the ongoing process of capitalist urbanization. A precarious balance among disparate political, economic, and cultural understandings of urban water supply systems has begun to disintegrate. The creation of metropolitan nature necessitated immense technical and organizational ingenuity in order to link the hydrological cycle of upstate New York to a multiplicity of private spaces within the city. The experience of the last twenty years has revealed how the prospect of a disintegrating and contaminated water system has exposed deep anxieties over the state of the public realm.

Building the Sidewalls, Open Cut No. 9. *From* Report to the Aqueduct Commissioners 1887, *City of New York*.

NOTES

1. Jean Gottmann, *Megalopolis: The Urbanized Northeastern Seaboard of the United States* (Cambridge, Mass.: MIT Press, 1961), 730.
2. In addition to the official opening ceremony, the new water tunnel has also been honored by an Obie-winning multimedia art performance dedicated to the twenty-four workers known as "sandhogs" who died during its construction. See Marty Pottenger, "CWT #3: Making City Water Tunnel #3," *High Performance* (Spring 1997), 6–14.
3. Nelson M. Blake, *Water for the Cities: A History of the Urban Water Supply Problem in the United States* (Syracuse, N.Y.: Syracuse University Press, 1956), 1.
4. Jean-Pierre Goubert, *The Conquest of Water: The Advent of Health in the Industrial Age*, trans. A. Wilson (1986; Cambridge: Polity Press, 1989), 259.
5. William Morrish, "The Urban Spring: Formalizing the Water System of Los Angeles," *Modulus: University of Virginia Architectural Review* 17 (1984): 45–83.
6. On the early history of New York City's water supply see George E. Hill and G. E. Waring, *Old Wells and Water-Courses of the Island of Manhattan* (New York: Knickerbocker Press, 1897); Irving V. A. Huie, "New York City's Water Supply and Its Future," *Municipal Engineers Journal* 37 (1951): 93–112; Gerard T. Koeppel, *Water for Gotham: A History* (Princeton, N.J.: Princeton University Press, 2000); Edward Wegmann, *The Water Supply of the City of New York, 1658–1895* (New York: John Wiley & Sons, 1896); Charles H. Weidner, *Water for a City: A History of New York City's Problem from the Beginning to the Delaware River System* (New Brunswick, N.J.: Rutgers University Press, 1974); Diane Galusha, *Liquid Assets: A History of New York City's Water System* (Fleischmanns, N.Y.: Purple Mountain Press, 2002).
7. Peter Kalm, *The America of 1750: Peter Kalm's Travels in North America*, ed. Adolph B. Benson (1770; New York: Wilson-Erickson, 1937), 133.
8. *New York Journal*, cited in Eugene Moehring, *Public Works and the Patterns of Urban Real Estate Growth in Manhattan, 1835–1894* (New York: Arno Press, 1981), 25.
9. Michael A. Mikkelson, "A Review of the History of Real Estate on Manhattan Island," in Real Estate Record Association, *A History of Real Estate, Building and Architecture in New York City* (New York: Record and Guide, 1898), 1–129.
10. See Alexander F. Vache, *Letters on Yellow Fever, Cholera and Quarantine; Addressed to the Legislature of the State of New York* (New York: McSpedon and Baker, 1852).
11. The relatively late incidence of cholera can be related to the increasing integration of the city into transatlantic trade and travel with Europe, which also fostered the racialization of the disease in the popular imagination. It was not until Robert Koch's discovery of the bacterium *Vibrio cholerae* in 1883 that earlier moral and miasmic conceptions of the disease began to wane. The historian Charles Rosenberg writes: "Cholera, a scourge of the sinful to many Americans in 1832, had, by 1866, become the consequence of remediable faults in sanitation." Charles E. Rosenberg, *The Cholera Years: The United States in 1832, 1849, and 1866* (Chicago: University of Chicago Press, 1962), 5. See also Leona Baumgartner, "One Hundred Years of Health: New York City, 1866–1966," *Bulletin of the New York Academy of Medicine* 45 (1969): 555–586; Gretchen A. Condran, "Changing Patterns of Epidemic Disease in New York City," and Alan M. Kraut, "Plagues and Prejudice: Nativism's Construction of Disease in Nineteenth- and Twentieth-Century New York City," both in David Rosner, ed., *Hives of Sickness: Public Health and Epidemics in New York City* (New Brunswick, N.J.: Rutgers University Press, 1995), 27–41, 65–90. Like cholera, yellow fever was spread in the late eighteenth and early nineteenth century by ships from areas where the disease was endemic, such as the West Indies and South America. Effective public health measures succeeded in bringing the disease under control, though it continued to affect southern US ports for decades. On broader debates concerning nineteenth-century disease and urbanization see, for example, Richard J. Evans, *Death in Hamburg: Society and Politics in the Cholera Years, 1830–1910* (Oxford: Oxford University Press, 1987); Richard J. Evans, "Epidemics and Revolutions: Cholera in Nineteenth-Century Europe," in T. Ranger and P. Slack, eds., *Epidemics and Ideas: Essays on the Historical Perception of Pestilence* (Cambridge: Cambridge University Press, 1992), 149–173; J. N. Hays, *The Burdens of Disease: Epidemics and Human Response in Western History* (New Brunswick, N.J.: Rutgers University Press, 1998); Gerry Kearns, "Biology, Class and the Urban Penalty," in G. Kearns and W. J. Withers, eds., *Urbanising Britain: Essays on Class and Community in the Nineteenth Century* (Cambridge: Cambridge University Press, 1991), 12–30.
12. John W. Francis, *Letter on the Cholera Asphyxia, Now Prevailing in the City of New-York* (New York: George P. Scott, 1832), 29.
13. Ibid.
14. Ibid.
15. New York City Board of Health, *Report of the Proceedings of the Sanatory Committee of the Board of Health, in Relation to the Cholera, As It Prevailed in New York in 1849* (New York: McSpedon and Baker, 1849), 11–12.
16. *The Cholera Bulletin*, intro. Charles E. Rosenberg (1832; New York: Arno Press, 1972). See also Dudley Atkins, ed., *Reports of Hospital Physicians and Other Documents in Relation to the Epidemic Cholera of 1832* (New York: Carvill, 1832).
17. Martyn Paine, *Letters on the Cholera Asphyxia, As It Appeared in the City of New-York* (New York: Collins and Hannay, 1832). See also David Meredith Reese, *Plain and Practical Treatise on the Epidemic Cholera, As It Prevailed in the City of New York, in the Summer of 1832* (New York: Conner and Cooke, 1833).
18. The Common Council had been created in 1686 as the city's legislative body.
19. Huie, "New York City's Water Supply and Its Future." In 1827 there were the last recorded attempts to gain water by drilling into Manhattan Island, in yet another ill-fated private intervention under the newly formed New York Well Company. See John Duffy, *A History of Public Health in New York City, 1625–1866* (New York: Russell Sage Foundation, 1968), 391.
20. New York State, *Act of Incorporation of the Manhattan Company* (New York: George Forman, 1799). The full text of the critical passage reads: "the said company shall, within ten years from the passing of this act, furnish and continue a supply of pure and wholesome water.... And it be further enacted, that it shall and may be lawful for the said company, to employ all such surplus capital as may belong or accrue to the said company in the purchase of public or other stock, or in any other monied transactions or operations, not inconsistent with the constitution and laws of this State or of the United States, for the sole benefit of the said company." (p. 11).
21. Duffy, *A history of Public Health in New York City, 1625–1866*, 28.
22. Ibid., 30.
23. New York Common Council, *Extract from the Minutes of the Common Council in Relation to the Manhattan Company* (New York: Childs & Devoe, 1835).
24. See Letty D. Anderson, "The Diffusion of Technology in the Nineteenth-Century American City: Municipal Water Supply Investments" (PhD dissertation, Northwestern University, 1980); Charles D. Jacobson and Joel A. Tarr, "The Development of Water Works in the United States," *Rassegna: Themes in Architecture* 57 (1994): 37–41; Maureen Ogle, "Water Supply, Waste Disposal, and the Culture of Privatism in the Mid-Nineteenth-Century American City," *Journal of Urban History* 25 (1999): 231–347; Robert Thorne, "The Hidden Iceberg of Architectural History," *Newsletter of the Society of Architectural Historians of Great Britain* 65 (1998): 8–9.

25. On the history of US water systems see Ellis L. Armstrong, Michael C. Robinson, and Suellen M. Hoy, eds., *History of Public Works in the United States, 1776–1976* (Chicago: American Public Works Association, 1976); Moses N. Baker, ed., *The Manual of American Water Works* (New York: Engineering News, 1889); Moses N. Baker, *The Quest for Pure Water: The History of Water Purification from the Earliest Records to the Twentieth Century* (Washington, D.C.: American Water Works Association, 1949); J. J. R. Croes, *Statistical Tables from the History and Statistics of American Water Works* (New York: Engineering News, 1883); Jacobson and Tarr, "The Development of Water Works in the United States"; Blake, *Water for the Cities*; J. Michael La Nier, "Historical Development of Municipal Water Systems in the United States, 1776–1976," *Journal of the American Water Works Association* (April 1976), 173–180; Abel Wolman, "75 years of Improvement in Water Supply Quality," *Journal of the American Water Works Association* 48 (1956): 905–914; Abel Wolman, "Status of Water Resources Use, Control, and Planning in the United States," *Journal of the American Water Works Association* 56 (1963): 1253–1272; Donald J. Pisani, "Beyond the Hundredth Meridian: Nationalizing the History of Water in the United States," *Environmental History* 4 (2000): 466–482.
26. Huie, "New York City's Water Supply and Its Future"; New York Committee on Fire and Water, *Report of the Board of Aldermen, Relative to Introducing into the City of New York a Supply of Pure and Wholesome Water* (28 December 1831).
27. On public health in nineteenth-century Philadelphia see, for example, Michael P. McCarthy, *Typhoid and the Politics of Public Health in Nineteenth-Century Philadelphia* (Philadelphia: American Philosophical Society, 1987); Nicholas B. Wainwright, ed., *A Philadelphia Perspective: the Diary of Sidney George Fisher Covering the Years 1834–1871* (Philadelphia: Historical Society of Philadelphia, 1967); Sam Bass Warner, *The Private City: Philadelphia in Three Periods of Its Growth* (Philadelphia: University of Pennsylvania Press, 1968).
28. See Donald J. Cannon, "Firefighting," in Kenneth T. Jackson, ed., *The Encyclopedia of New York* (New Haven: Yale University Press, 1995), 408–412. Interestingly, the last great conflagration of 1835 actually accelerated the speculative boom of the 1830s through a kind of proto-Haussmannization of the downtown business district, which could now be completely rebuilt.
29. Moehring, *Public Works and the Patterns of Urban Real Estate Growth in Manhattan*.
30. City bond issues were essential because the private sector was incapable of raising capital on the scale that was required. See Jacobson and Tarr, "The Development of Water Works in the United States"; Laura Rosen, "New York Builds in Hard Times," *Livable City* 16 (1992): 2–4.
31. Cited in Moehring, *Public Works and the Patterns of Urban Real Estate Growth in Manhattan*, 39. See also Roger Panetta, "The Croton Aqueduct and Suburbanization of Westchester," in Hudson River Museum of Westchester, *The Old Croton Aqueduct: Rural Resources Meet Urban Needs* (Yonkers, N.Y.: Hudson River Museum of Westchester, 1992), 41–48.
32. Cited in Moehring, *Public Works and the Patterns of Urban Real Estate Growth in Manhattan*, 51.
33. Philip Hone writing in 1842, cited in I. N. Phelps Stokes, *The Iconography of Manhattan Island* (1917; New York: Arno Press, 1967), p. 1777.
34. See *The Diary of Philip Hone*, ed. B. Tuckerman, vol. 2, 1828–1851 (New York: Dodd, Mead & Company, 1889).
35. Vittorio Gregotti, editorial in *Rassegna: Themes in Architecture* 57 (1994): 5.
36. For six years, four thousand mostly Irish workers had worked on the largest public works project in the city's history up to that time. See Charles King, *A Memoir of the Construction, Cost, and Capacity of the Croton Aqueduct* (New York: Charles King, 1843); John B. Jervis, *Description of the Croton Aqueduct* (New York: Slamm and Guion, 1842); "The Boston Miscellanies: The Croton Aqueduct," in *Being a Collection of Useful and Entertaining Articles on Various Subjects* (Boston: Bradbury & Guild, 1850), 28–36. More recent sources include Larry D. Lankton, *The "Practicable" Engineer: John B. Jervis and the Old Croton Aqueduct*, Essays on Public Works History, no. 5. (Chicago: Public Works Historical Society, 1977); F. Daniel Larkin, *John B. Jervis: An American Engineering Pioneer* (Arnes: Iowa State University Press, 1990); George H. Rappole, "The Old Croton Aqueduct," *IA: Journal of the Society for Industrial Archaeology* 1 (1978): 15–25; Hudson River Museum of Westchester, *The Old Croton Aqueduct*.
37. Jervis's experience on the Erie Canal is significant because this was not only one of the earliest examples of large-scale public works in North America but also accelerated the trade-based regional interconnectedness that underpinned the rapid expansion of nineteenth-century New York.
38. David E. Nye, *American Technological Sublime* (Cambridge, Mass.: MIT Press, 1994), 1. On the iconography of nineteenth-century water engineering and changing conceptions of American landscape, see Patrick McGreevy, "Imagining the Future at Niagara Falls," *Annals of the Association of American Geographers* 77 (1987): 48–62, and William Irwin, *The New Niagara: Tourism, Technology, and the Landscape of Niagara Falls, 1776–1917* (University Park: Pennsylvania State University Press, 1996). On the cultural history of American technology see also Joseph J. Corn, ed., *Imagining Tomorrow: History, Technology and the American Future* (Cambridge, Mass.: MIT Press, 1986), and Leo Marx, *The Pilot and the Passenger: Essays on Literature, Technology and Culture in the United States* (Oxford: Oxford University Press, 1988).
39. See Laura Vookles Hardin, "Celebrating the Aqueduct: Pastoral and Urban Ideals," in Hudson River Museum of Westchester, *The Old Croton Aqueduct*, 49–56. See also Marvin Fisher, *Workshops in the Wilderness: The European Response to American Industrialization, 1830–1860* (New York: Oxford University Press, 1967); John F. Kasson, *Civilizing the Machine: Technology and Republican Values in America, 1776–1900* (New York: Grossman, 1976); and Carroll Pursell, *The Machine in America: A Social History of Technology* (Baltimore: Johns Hopkins University Press, 1995).
40. George Templeton Strong, cited in Gerard Koeppel, "A Struggle for Water," *Invention and Technology* 9 (1994): 18–31.
41. See Joanne Abel Goldman, *Building New York's Sewers: Developing Mechanisms of Urban Management* (West Lafayette, Ind.: Purdue University Press, 1997), 51.
42. See Moehring, *Public Works and the Patterns of Urban Real Estate Growth in Manhattan*.
43. Sam Bass Warner makes a similar observation with respect to nineteenth-century Philadelphia: "Fears of epidemics had created the water system, but once this fear had abated, little or no public support remained to bring the benefits of the new technology to those who could not afford them." Warner, *The Private City*, 3. On the significance of municipal achievement for the legitimacy of urban government, see Jon C. Teaford, *The Unheralded Triumph: City Government in America, 1870–1900* (Baltimore: Johns Hopkins University Press, 1984), and Raymond Mohl, *The Making of Urban America* (Wilmington, Del.: Scholarly Resources, 1988). The fetishistic nature of urban infrastructure as a means to mask unchanged social relations is tackled in Maria Kaïka and Erik Swyngedouw, "Fetishising the Modern City: the Phantasmagoria of Urban Technological Networks," *International Journal of Urban and Regional Research* 24 (2000): 120–138. On the politics of public health reform in American cities see also Barbara Gutmann Rosenkrantz, *Public Health and the State: Changing Views in Massachusetts, 1842–1936* (Cambridge, Mass.: Harvard University Press, 1972); Judith Walzer

Leavitt, *The Healthiest City: Milwaukee and the Politics of Health Reform* (Princeton, N.J.: Princeton University Press, 1982); and Stuart Galishoof, *Newark: The Nation's Unhealthiest City, 1832–1895* (New Brunswick, N.J.: Rutgers University Press, 1988).
44. See André Guillerme, "Water for the City," *Rassegna: Themes in Architecture* 57 (1994): 6–21.
45. Joel Shew, *The Water-Cure Manual* (New York: La Morte Barney, 1847), 2. See also John G. Coffin, *Discourses on Cold and Warm Bathing; with Remarks on the Effects of Drinking Cold Water in Warm Weather*, 2d ed. (Boston: Cummings and Hilliard, 1826).
46. Sigfried Giedion, *Mechanization Takes Command: A Contribution to Anonymous History* (New York: Oxford University Press, 1948), 684.
47. See Richard L. Bushman and Claudia L. Bushman, "The Early History of Cleanliness in America," *Journal of American History* 74 (1988): 1213–1238; Maureen Ogle, "Domestic Reform and American Household Plumbing, 1840–1870," *Winterthur Portfolio* 28 (1993): 33–58; Maureen Ogle, *All the Modern Conveniences: American Household Plumbing, 1840–1890* (Baltimore: Johns Hopkins University Press, 1996); Richard L. Bushman, *The Refinement of America: Persons, Houses, Cities* (New York: Knopf, 1992); Suellen Hoy, *Chasing Dirt: The American Pursuit of Cleanliness* (New York: Oxford University Press, 1995).
48. Alain Corbin, *The Foul and the Fragrant: Odor and the French Social Imagination* (1982; Cambridge, Mass.: Harvard University Press, 1986). See also Pierre Bourdieu, *Distinction: A Social Critique of the Judgement of Taste* (1979; London: Routledge, 1986).
49. See, for example, Lawrence Wright, *Clean and Decent: The Fascinating History of the Bathroom and the Water Closet* (New York: Viking Press, 1960); Stuart M. Blumin, *The Emergence of the Middle Class: Social Experience in the American City, 1760–1900* (Cambridge: Cambridge University Press, 1989); Nancy Tomes, "The Private Side of Public Health: Sanitary Science, Domestic Hygiene and the Germ Theory, 1870–1900," *Bulletin of the History of Medicine* 64 (1990): 467–480.
50. William Paul Gerhard, *The Drainage of a House* (Boston: Rand Avery, 1888); Giedion, *Mechanization Takes Command*; New York City Committee of Seventy, *Preliminary Report of the Sub-Committee on Baths and Lavatories*, 1895. Although water became a fashionable cultural commodity, the adoption of bathrooms in every home did not occur until the twentieth century. The standard bathroom design, for example, did not emerge until the mass production of enameled fixtures in the 1920s and the spread of the "American Compact Bathroom" with its origins in hotel architecture. See also Susan J. Kleinberg, "Technology and Women's Work: The Lives of Working Class Women in Pittsburgh, 1870–1900," in Martha Moore Trescott, ed., *Dynamos and Virgins Revisited* (Metuchen, N.J.: Scarecrow Press, 1979), 185–204; John Duffy, *A History of Public Health in New York City, 1866–1966* (New York: Russell Sage Foundation, 1974), 44; Marilyn Thornton Williams, "New York City's Public Baths: A Case Study in Urban Progressive Reform," *Journal of Urban History* 7 (1980): 49–81.
51. See Anthony Jackson, *A Place Called Home: A History of Low-Cost Housing in Manhattan* (Cambridge, Mass.: MIT Press, 1976); David Glassberg, "The Public Bath Movement in America," *American Studies* 20 (1979): 5–21; Ogle, "Water Supply, Waste Disposal, and the Culture of Privatism"; Richard Plunz, *A History of Housing in New York City: Dwelling Type and Social Change in the American Metropolis* (New York: Columbia University Press, 1990).
52. George E. Waring, "The Disposal of a City's Waste," *North American Review* (July 1895), 49.
53. New York City Metropolitan Sewerage Commission, *Sewerage and Sewage Disposal in the Metropolitan District of New York and New Jersey* (New York: Martin B. Brown, 1910).
54. The sanitary inspector Robert Newman, cited in James E. Serrell, *Compilation of Facts Representing the Present Condition of the Sewers and Their Deposits in the City of New York* (New York: Bergen & Tripp, 1866), 10.
55. Gail Caskey Winkler and Charles E. Fischer III, *The Well-Appointed Bath: Authentic Plans and Fixtures from the Early 1900s* (Washington, D.C.: Preservation Press/National Trust for Historic Preservation, 1989).
56. John R. Freeman, *Report upon New York's Water Supply with Particular Reference to the Need of Procuring Additional Sources and Their Probable Cost with Works Constructed under Municipal Ownership* (New York: Martin B. Brown, 1900). New York's ascendancy was clearly short-lived: by 1855, for example, New York had only 1,361 baths for its population of 629,904 compared with 3,521 for a population of nearly 340,000 in Philadelphia. See Edgar W. Martin, *The Standard of Living in 1860* (Chicago: University of Chicago Press, 1942).
57. See Bushman and Bushman, "The Early History of Cleanliness in America"; Ellen Lupton and J. Abbott Miller, *The Bathroom, the Kitchen and the Aesthetics of Waste: A Process of Elimination* (New York: Kiosk, 1992).
58. Weidner, *Water for a City*, 115. See also Duffy, *A History of Public Health in New York City, 1866–1966*, 77.
59. James C. Bayles, *Pipe Gallery Experience* (New York: Martin B. Brown, 1903), 3.
60. Elisha Harris, *Cholera Prevention: Examples and Practice and a Note on the Present Aspects of the Epidemic* (New York: Appleton & Company, 1867).
61. Gerald R. Iwan, "Drinking Water Quality Concerns of New York City, Past and Present," *Annals of the New York Academy of Sciences* 502 (1987): 183–204. The first water quality laboratories for an American city had been established in Boston in 1889. On the history of water pollution in upstate New York see also G. Harris and S. Wilson, "Water Pollution in the Adirondack Mountains: Scientific Research and Governmental Response, 1890–1930," *Environmental History Review* 17 (1993): 47–81.
62. Arnold H. Ruge to William L. Strong, "Rational Improvements to Purify the Water Supply of the City of New York" (17 December 1894). Papers held by the New-York Historical Society.
63. Other changes include a reforestation program and improved policing of the city's watershed. See Frank E. Hale, "The Development of the Sanitary Safeguards of the New York City Water Supply and Its Relation to Typhoid Fever and Other Diseases," *Municipal Engineers Journal* 7 (1921): 30–61. As a result of these different initiatives, typhoid death rates in Manhattan fell from around 20 per 100,000 population in 1901 to under 2 by 1919.
64. On the emergence of urban planning in American cities see, for example, M. Christine Boyer, *Dreaming the Rational City: The Myth of American City Planning* (Cambridge, Mass.: MIT Press, 1983); John D. Fairfield, "The Scientific Management of Urban Space: Professional City Planning and the Legacy of Progressive Reform," *Journal of Urban History* 20 (1994): 179–204; Nelson P. Lewis, *The Planning of the Modern City* (New York: Wiley, 1916); Stanley K. Schultz and Clay McShane, "To Engineer the Metropolis: Sewers, Sanitation, and City Planning in Late Nineteenth-Century America," *Journal of American History* 65 (1978): 389–411; Joel A. Tarr, *The Search for the Ultimate Sink: Urban Pollution in Historical Perspective* (Akron, Ohio: University of Akron Press, 1996).
65. George T. Hammond, "City Planning and the Engineer," *Municipal Engineers Journal* 2 (1916): 368, 338.
66. Mike Davis, *Prisoners of the American Dream: Politics and Economy in the History of the US Working Class* (London: Verso, 1986), 120. See also Andrew Jamison, "American Anxieties: Technology and the Reshaping of Republican Values," in Mikael Hård and Andrew Jamison, eds., *The Intellectual Appropriation of Technology: Discourses on Modernity, 1900–1939* (Cambridge, Mass.: MIT Press, 1998), 69–100.
67. On the evolution and diversity of American engineering traditions see David Hovey Calhoun, *The*

American Civil Engineer: Origins and Conflict (Cambridge, Mass.: MIT Press, 1960); Monte A. Calvert, *The Mechanical Engineer in America, 1830–1910* (Baltimore: Johns Hopkins University Press, 1967); R. S. Kirby, S. Withington, A. B. Darling, and F. G. Kilfour, *Engineering in History* (New York: McGraw-Hill, 1956); Donald C. Jackson, "Engineering in the Progressive Era: A New Look at Frederick Haynes Newell and the US Reclamation Service," *Technology and Culture* 34 (1993): 539–574; Peter Meiksins, "The 'Revolt of the Engineers' Reconsidered," in Terry S. Reynolds, ed., *The Engineer in America: A Historical Anthology from Technology and Culture* (Chicago: University of Chicago Press, 1991), 399–426; R. H. Merritt, *Engineering in American Society, 1850–1875* (Lexington: University of Kentucky Press, 1969); and Tom F. Peters, *Building the Nineteenth Century* (Cambridge, Mass.: MIT Press, 1996).

68. The tension between technical expertise and political power can be illustrated by the evolving relationship between New York City government and New York State-based interests. In the period between 1849 and 1857, for example, a series of city charter revisions undertaken by the State Legislature sought to strengthen the power of technical elites over newly emerging city-based political machines. The 1857 charter revision was especially significant in increasing the role of engineers and technical experts in urban management. In 1870, however, a new city charter firmly reasserted the power of city-based political interests and established a powerful new Department of Public Works, which was to become a pivotal focus of power and patronage in the Tweed era of New York City governance. In the wake of the 1873 financial panic there was a renewed emphasis on investment in public works, and the transformation of New York's physical environment gathered pace. See Goldman, *Building New York's Sewers*; Moehring, *Public Works and the Patterns of Urban Real Estate Growth in Manhattan*, 39; Schultz and McShane, "To Engineer the Metropolis." As the cities grew, public contracts multiplied for water, gas, electricity, street railroads, and telephones, which helps to explain the rationale of greater autonomy of civil institutions from the patronage of local government. See Boyer, *Dreaming the Rational City*; Sarah S. Elkind, "Building a Better Jungle: Anti-Urban Sentiment, Public Works, and Political Reform in American Cities, 1880–1930," *Journal of Urban History* 24 (1997): 53–88.
69. Access to Croton water had proved a significant pretext for these outer boroughs' voting to join New York City. The water shortages facing Brooklyn in the late nineteenth century provided further impetus for the expansion of the Croton system and a new emphasis on regional water resources planning. See Brooklyn Division of Water Supply and I. M. deVarona, *History and Description of the Water Supply of the City of Brooklyn* (Brooklyn, N.Y.: Brooklyn Commissioner of City Works, 1896); Theodore Weston, *A Report on the Extent and Character of the District Supplying Water to the City of Brooklyn* (Brooklyn, N.Y.: D. Van Nostrand, 1861). A systematic analysis of the city's water needs was developed by 1903 by the Hering-Burr-Freeman Commission, which laid the scientific basis for the expansion of the water system into the Catskill Mountains. The political impetus for the completion of the project was also bolstered by the drought of 1910, the worst since 1826.
70. See Board of Water Supply of the City of New York, *A General Description of the Catskill Water Supply and of the Project for an Additional Supply from the Delaware River Watershed and the Rondout Creek* (1940); J. Paul Sehgal, "New York's Water Problems," *Municipal Engineers Journal* 79 (1991): 24–34.
71. Gilbert J. Fowler, cited in Metropolitan Sewerage Commission of New York, *Preliminary Reports on the Disposal of New York's Sewage VII: Critical Reports of Dr Gilbert J. Fowler of Manchester, England, and Mr John D. Watson of Birmingham, England, on the Projects of the Metropolitan Sewerage Commission with Special Reference to the Plans Proposed for the Lower Hudson, Lower East River and Bay Division* (1913), 33.
72. Weidner, *Water for a City*, 140.
73. See Silas B. Dutcher to Bird S. Coler, 5 August 1899, papers held by the New-York Historical Society. This letter reveals the failure of the Ramapo Water Company to divulge requested technical and financial information to the city's comptroller.
74. T. Hochlerner, "Distribution System for New York," *Water Works Engineering* 95 (1942): 1262–1266.
75. See, for example, William L. Kahrl, *Water and Power: The Conflict over Los Angeles' Water Supply in the Owens Valley* (Berkeley: University of California Press, 1982); R. Lowitt, "The Hetch-Hetchy Controversy," *California History* 74 (1995): 190–203; Robert A. Sauder, *The Lost Frontier: Water Diversion and the Destruction of Owens Valley Agriculture* (Tuscon: University of Arizona Press, 1993); John Walton, *Western Times and Water Wars: State, Culture, and Rebellion in California* (Berkeley: University of California Press, 1992); Weidner, *Water for a City*, 231.
76. Weidner, *Water for a City*, 241, 263.
77. Ibid., 191, 272.
78. Ibid., 271. See also E. J. Clark, "The New York City Water Shortage," *Municipal Engineers Journal* 36 (1950): 92–104; Anastasia Van Burkalow, "The Geography of New York City's Water Supply: a Study of Interactions," *Geographical Review* 49 (1959): 369–386.
79. *New York Herald*, cited in Weidner, *Water for a City*, 271. The commemorative medal of the mayor's Catskill Celebration Committee described the new water system as "an achievement of civic spirit, scientific genius, and faithful labor." See Edward E. Hall, *Water for New York City* (1917; Saugerties, N.Y.: Hope Farm Press, 1993), 122.
80. Weidner, *Water for a City*, 290.
81. Board of Water Supply of the City of New York, *A General Description of the Catskill Water Supply*, 1940, papers held by the New-York Historical Society.
82. Charles M. Clark, "Development of the Catskill Supply," *Water Works Engineering* 95 (1942): 1268–1269; Charles M. Clark, "History of the Delaware Supply," *Water Works Engineering* 95 (1942): 1276–1277; Fiorello La Guardia, "A Letter from Mayor La Guardia," *Water Works Engineering* 95 (1942): 1256.
83. Rexford G. Tugwell, "San Francisco as Seen from New York," *National Conference on Planning* (proceedings of the conference held at San Francisco, 8–11 July 1940), 182–188. See also Thomas Kessner, *Fiorello H. La Guardia and the Making of Modern New York* (New York: McGraw-Hill, 1989); William Stanley Parker, "Public Works: The Future Share of Federal and Non-Federal Agencies," *National Conference on Planning* (proceedings of the conference held at Boston, Massachusetts, 15–17 May 1939).
84. "New York City's Gigantic Thirst," *New York Herald Tribune*, 15 August 1959.
85. Edward J. Clark, "New York City Water Supply System, Unusual Occurrences and Unique Problems," *Municipal Engineers Journal* 34 (1948): 103–115.
86. "Mayor Asks Curb on Use of Water," *New York Times*, 15 October 1963.
87. Abel Wolman, "The Metabolism of Cities," *Scientific American* 213 (1965): 180.
88. Lewis Mumford, "The Sky Line: The Architecture of Power," *New Yorker* (7 June 1941), 58–60. See also Aaron Betsky, "Measured Immensity: Hoover Dam at Fifty," *Progressive Architecture* 66 (1985): 38; Brian Black, "Authority in the Valley: TVA in Wild River and the Popular Media, 1930–1940," *Journal of American Culture* 18 (1995): 1–14; Margot W. Garcia, "An Everlasting Monument: The Building of Roosevelt Dam," *Triglyph: A Southwestern Journal of Architecture and Environmental Design* 5 (1993): 34–46; Jean-François Lejeune, "Democratic Pyramids: The Works of the Tennessee Valley Authority," *Rassegna: Themes in Architecture* 63 (1995): 46–57; Marian Moffett and Lawrence Wodehouse, "Noble Structures Set in Handsome Parks: Public Architecture of the TVA," *Modulus: University of Virginia Architectural Review* 17

(1984): 74–83; Richard Guy Wilson, "Massive Deco Monument: The Enduring Strength of Boulder (Hoover) Dam," *AIA: Architecture* (December 1983), 45–47; Richard Guy Wilson, "Machine-Age Iconography in the American West: The Design of Hoover Dam," *Pacific Historical Review* 54 (1985): 463–497.

89. See Susan Fainstein and Norman Fainstein, "Technology, the New International Division of Labor, and Location: Continuities and Disjunctures," in Robert Beauregard, ed., *Economic Restructuring and Political Response* (Newbury Park, Calif.: Sage Publications, 1989).
90. "How to Sell the World's Water Industry," *Financial Times*, 2 October 1997.
91. "Controversy Hits Manila Water Privatisation Plan," *Financial Times*, 17 January 1997; "Water: Trouble in the Pipeline," *Financial Times*, 14 September 1998; "Doubts over Brazilian Water Privatisation," *Financial Times*, 13 November 1998; "Anger in the Andes," *Financial Times*, 4 July 2000; David Saurí and Francisco Manuel Muñoz, "The Limits of Ecological Modernization in the Multiplied City: Equity and Conflict over Water Costs in the Metropolitan Area of Barcelona," in Erik Swyngedouw, Leandro del Moral, and Grigoris Kafkalas, eds., *Sustainability, Risk and Nature: the Political Ecology of Water in Advanced Societies* (Oxford: Oxford Centre for Water Research, 1999), 195–201.
92. See M. J. Pfeffer and J. M. Stycos, *Watershed Views: a Public Opinion Survey on the New York City Watershed* (Ithaca, N.Y.: Department of Rural Sociology and Population and Development Program, Cornell University, 1994); K. A. Stave, "Resource Conflict in the New York City Catskill Watersheds: A Case for Expanding the Scope of Water Resource Management," in L. H. Austin, ed., *Water in the 21st Century: Conservation, Demand and Supply* (Herndon, Va.: Proceedings of the American Water Works Association Annual Spring Symposium, 1995), 61–68.
93. See David C. Major, "Urban Water Supply and Global Environmental Change: The Water Supply System of New York City," in R. Herrmann, ed., *Proceedings of the 28th Annual American Water Resources Association Conference and Symposium: Managing Resources during Global Change* (Bethesda, Md.: American Water Resources Association, 1992), 377–385; "Getting Ready for a Hotter, Wetter Future," *New York Times*, 1 December 1997; "Report Warns New York of Perils of Global Warming," *New York Times*, 30 June 1999; "Water Supply Also Affected by Global Warming," *Village Voice* (3 August 1993).
94. Arjun Appadurai, *Modernity at Large: Cultural Dimensions of Globalization* (Minneapolis: University of Minnesota Press, 1996), 19. For starkly different views of the relationship between politics and modernity see Jürgen Habermas, *The Philosophical Discourse of Modernity: Twelve Lectures* (1985; Cambridge: Polity Press, 1987); Eric Hobsbawm, *Age of Extremes: The Short Twentieth Century, 1914–1991* (London: Michael Joseph, 1994); and E. M. Wood, *Democracy against Capitalism: Renewing Historical Materialism* (Cambridge: Cambridge University Press, 1995).
95. See the critique of Donald Worster's conspiratorial view of urban water demands developed by Douglas Edward Kupel, "Urban Water in the Arid West: Municipal Water and Sewer Utilities in Phoenix, Arizona" (PhD dissertation, Arizona State University, 1995).

PLATE 1
Old Croton Aqueduct, Ossining, New York, 1997

PLATE 2
High Bridge, Bronx and New York, New York, 2000

PLATE 3
High Bridge, Bronx and New York, New York, 1995

PLATE 4
Old Croton Aqueduct, Bronx, New York, 2000

PLATE 5
High Bridge Tower, New York, New York, 1995

PLATE 6
Valve, Old Croton Aqueduct, Ossining, New York, 1997

PLATE 7
Gatehouse, Old Croton Aqueduct, Ossining, New York, 1997

PLATE 8

Dam and Spillway, Cross River Reservoir, Westchester County, New York, 2000

PLATE 9
Spillway, Cross River Reservoir, Westchester County, New York, 2000

PLATE 10
Stilling Basin, Cross River Reservoir, Westchester County, New York, 2000

PLATE 11
Gatehouse and Spillway, West Branch Reservoir, Putnam County, New York, 1995

PLATE 12

Gatehouse, West Branch Reservoir, Putnam County, New York, 1997

PLATE 13
Dam and Spillway, Titicus Reservoir, Westchester County, New York, 2000

PLATE 14
Spillway, East Branch Reservoir, Putnam County, New York, 1999

PLATE 15

Spillway, Croton Falls Diverting Reservoir, Putnam County, New York, 1994

PLATE 16
Croton Falls Diverting Reservoir, Putnam County, New York, 1999

PLATE 17
Connecting Channel Gatehouse, Putnam County, New York, 1995

PLATE 18
Connecting Channel Gatehouse, Putnam County, New York, 2000

PLATE 19
Dam and Spillway, Croton Falls Reservoir, Putnam County, New York, 1995

PLATE 20

Spillway, Croton Falls Reservoir, Putnam County, New York, 2000

PLATE 21
New Croton Dam, Westchester County, New York, 1999

PLATE 22
New Croton Dam, Westchester County, New York, 1999

PLATE 23
Spillway, New Croton Dam, Westchester County, New York, 1999

PLATE 24
Stilling Basin, New Croton Dam, Westchester County, New York, 1999

PLATE 25
Upper Gatehouse, New Croton Dam, Westchester County, New York, 2000

PLATE 26
Lower Gatehouse, New Croton Dam, Westchester County, New York, 2000

PLATE 27
New Croton Gatehouse, Westchester County, New York, 2000

PLATE 28
New Croton Gatehouse, Westchester County, New York, 2000

PLATE 29
New Croton Aqueduct Gatehouse, South Yonkers, New York, 1999

PLATE 30
Jerome Park Reservoir, Bronx, New York, 2000

PLATE 31
135th Street Gatehouse, New York, New York, 2001

PLATE 32
Gatehouse, Massapequa, New York, 1998

PLATE 33
Millburn Pumping Station, Freeport, New York, 1998

PLATE 34
Gatehouse, Wantagh, New York, 1997

PLATE 35
Upper Gate Chamber and Dividing Weir, Ashokan Reservoir, Ulster County, New York, 1999

PLATE 36
Dividing Weir, Ashokan Reservoir, Ulster County, New York, 1999

PLATE 37
Dividing Weir, Ashokan Reservoir, Ulster County, New York, 1999

PLATE 38

Upper Gate Chamber, Ashokan Reservoir, Ulster County, New York, 2000

PLATE 39

Upper Gate Chamber, Ashokan Reservoir, Ulster County, New York, 2000

PLATE 40
Olive Bridge Dam, Ashokan Reservoir, Ulster County, New York, 1999

PLATE 41
Stairway, Olive Bridge Dam, Ashokan Reservoir, Ulster County, New York, 1999

PLATE 42
Esopus Creek at Olive Bridge Dam, Ashokan Reservoir, Ulster County, New York, 2000

PLATE 43
Waste Weir, Ashokan Reservoir, Ulster County, New York, 1999

PLATE 44

Spillway, Neversink Reservoir, Sullivan County, New York, 1999

PLATE 45
Spillway, Neversink Reservoir, Sullivan County, New York, 2000

PLATE 46
Spillway, Neversink Reservoir, Sullivan County, New York, 1999

PLATE 47
Spillway, Neversink Reservoir, Sullivan County, New York, 2000

PLATE 48
Stilling Basin, Neversink Reservoir, Sullivan County, New York, 1999

PLATE 49
Spillway, Merriman Dam, Rondout Reservoir, Ulster County, New York, 1999

PLATE 50
Spillway, Downsville Dam, Pepacton Reservoir, Delaware County, New York, 1997

PLATE 51
Spillway, Downsville Dam, Pepacton Reservoir, Delaware County, New York, 1997

PLATE 52
Spillway Gatehouse, Downsville Dam, Pepacton Reservoir, Delaware County, New York, 1997

PLATE 53
Spillway, Stilesville Dam, Cannonsville Reservoir, Delaware County, New York, 1997

PLATE 54
Kensico Dam, Valhalla, New York, 1999

PLATE 55
Kensico Dam, Valhalla, New York, 1999

PLATE 56
Kensico Dam, Valhalla, New York, 1999

PLATE 57
Aerator Field, Kensico Reservoir, Valhalla, New York, 1994

PLATE 58
Delaware Aqueduct Influent, Kensico Reservoir, Valhalla, New York, 2000

PLATE 59
Dewatering Apparatus, Shaft 6, Delaware Aqueduct, Chelsea, New York, 2000

PLATE 60
Shaft 6, Delaware Aqueduct, Chelsea, New York, 2000

PLATE 61
Shaft 6, Delaware Aqueduct, Chelsea, New York, 2000

PLATE 62
Shaft 9, Delaware Aqueduct, West Branch Reservoir, Putnam County, New York, 1997

PLATE 63
Roof, Dewatering Shaft, Catskill Aqueduct, Westchester County, New York, 2000

PLATE 64

Gatehouse, Hillview Reservoir, Yonkers, New York, 2000

PLATE 65
Bypass Tunnel, Hillview Reservoir, Yonkers, New York, 2000

PLATE 66
Dewatering Apparatus, Shaft 21, City Tunnel No. 1, New York, New York, 2000

PLATE 67
Dewatering Apparatus, Shaft 21, City Tunnel No. 1, New York, New York, 2000

PLATE 68
Shaft 2b, City Tunnel No. 3, Bronx, New York, 1992

PLATE 69
Shaft 2b, City Tunnel No. 3, Bronx, New York, 1992

PLATE 70
Shaft 2b, City Tunnel No. 3, Bronx, New York, 1992

PLATE 71

Shaft 13b, City Tunnel No. 3, New York, New York, 1994

PLATE 72
Shaft 13b, City Tunnel No. 3, New York, New York, 1994

PLATE 73
Shaft 15b, City Tunnel No. 3, Roosevelt Island, New York, New York, 2000

PLATE 74
Shaft 15b, City Tunnel No. 3, Roosevelt Island, New York, New York, 2000

PLATE 75
City Tunnel No. 3, Queens, New York, 1997

PLATE 76
City Tunnel No. 3, Queens, New York, 1998

PLATE 77
City Tunnel No. 3, Queens, New York, 1998

PLATE 78
City Tunnel No. 3, Queens, New York, 1998

PLATE 79
City Tunnel No. 3, Queens, New York, 1997

PLATE 80
City Tunnel No. 3, Brooklyn, New York, 1998

PLATE 1

PLATE 2

PLATE 3

PLATE 4

PLATE 5

PLATE 6

PLATE 7

CATALOG

Croton System

PLATE 1 Old Croton Aqueduct, Ossining, New York, 1997
PLATE 2 High Bridge, Bronx and New York, New York, 2000
PLATE 3 High Bridge, Bronx and New York, New York, 1995
PLATE 4 Old Croton Aqueduct, Bronx, New York, 2000
PLATE 5 High Bridge Tower, New York, New York, 1995
PLATE 6 Valve, Old Croton Aqueduct, Ossining, New York, 1997
PLATE 7 Gatehouse, Old Croton Aqueduct, Ossining, New York, 1997

OLD CROTON AQUEDUCT

In the early nineteenth century, the connection between unsanitary water and disease epidemics had recently been discovered. At the same time, large fires plagued New York City. The city made plans to divert water from upstate, and in 1842, after many false starts, the Old Croton Aqueduct was completed, bringing a reliable source of water from the Croton Reservoir. When the water finally came, it allowed New York City to grow at a phenomenal rate.

The Croton system was designed and built by one of the country's great engineers, John P. Jervis, who was responsible for its surveying, design, and construction. Water was collected at the Croton Reservoir, created by the construction of the Croton Dam, then directed into the forty-mile Croton Aqueduct across the Harlem River to Manhattan via the High Bridge. (During the Civil War, there was fear of Confederate soldiers poisoning the water or attacking the aqueduct, and guards were stationed at High Bridge and along the aqueduct.) Finally, it came to the receiving reservoir in Central Park and the distributing reservoir at Forty-second Street (now the site of the New York Public Library, where some of the original walls of the reservoir were incorporated into the foundation). This reservoir was expected to serve all the city's needs for decades, but was soon insufficient, and additional reservoirs, including another in Central Park, had to be built.

Except for a small section that supplies Sing Sing Correctional Facility, the aqueduct has been out of service since 1958. Though much of the Old Croton Aqueduct has been removed from Manhattan, the rest of its route is largely intact. Most of the Westchester part of the aqueduct's route can be walked or biked, and parts of the Bronx route are still there. Along the way, several artifacts remain. There are gatehouses, aerators, and culverts, most dating back to the original construction. There may even be a few remnants in Central Park (the original reservoir there was replaced by the current reservoir, also currently off-line).

I was able to go inside the aqueduct in two places: Ossining, where it crosses a bridge over a small valley, and underneath the Jerome Park Reservoir, which was built between 1895 and 1905 for the New and Old Croton aqueducts. At Ossining, several incandescent lights have been installed, changing the appearance of the tunnel but making photography relatively easy. At Jerome Park, there were no lights, but I managed to light a small portion of the tunnel to photograph it.

HIGH BRIDGE

There was much controversy as to whether to build a high or low bridge or a tunnel to move the Croton water across the Harlem River. Although Jervis suggested a low bridge, the High Bridge finally won out. Even he understood that the larger structure would create a monument to the water system, and, in fact, many New Yorkers made day trips to see the bridge. Jervis worried that settling would cause instability in the bridge structure. He was correct, but it took much longer than he expected. Only now is the bridge, the city's oldest, in need of stabilization; many would like to see it opened to pedestrian traffic if it can be repaired.

HIGH BRIDGE TOWER

The High Bridge tower, along with a small reservoir next to it, was built between 1866 and 1870, so that the higher elevations of Manhattan could also receive water. Since the system was gravity fed, there was not enough "head" to bring the water up high enough to reach some people's homes. A pump station sent water into the tower, from where it could flow back down to the homes. There were other towers in Manhattan and Brooklyn for the same purpose, but this is the only one still standing.

PLATE 8

PLATE 9

PLATE 10

PLATE 11

PLATE 12

PLATE 13

Croton System

PLATE 8 Dam and Spillway, Cross River Reservoir, Westchester County, New York, 2000

PLATE 9 Spillway, Cross River Reservoir, Westchester County, New York, 2000

PLATE 10 Stilling Basin, Cross River Reservoir, Westchester County, New York, 2000

PLATE 11 Gatehouse and Spillway, West Branch Reservoir, Putnam County, New York, 1995

PLATE 12 Gatehouse, West Branch Reservoir, Putnam County, New York, 1997

PLATE 13 Dam and Spillway, Titicus Reservoir, Westchester County, New York, 2000

WEST BRANCH RESERVOIR

The West Branch Reservoir near Carmel in Putnam County was completed in 1896. When the Delaware Aqueduct was built, between 1939 and 1945, it was converted into a connection to allow the transfer of water between the Croton and Delaware systems. I went to see the dam several times. Once, I was asked by a DEP worker if I would like to go out on the frozen reservoir to get a different view of the gatehouse. I hesitated, but when I saw people driving on the ice in pickup trucks to get to their ice-fishing spots, I realized it was probably safe. However, I am not sure it is a good idea to allow trucks to drive on our drinking water, even when it is frozen.

CROSS RIVER RESERVOIR

Cross River Dam, located about a mile east of Katonah, was completed in 1908 after four years of construction. It was built with steel reinforcing rods, possibly the first use of this technique. In the 1990s the dam was upgraded to meet federal safety requirements, and a new spillway and stilling basin were installed to absorb some of the force of the water released from the reservoir, which otherwise would erode the stream bed. In some ways, the dam's design echoes that of the New Croton Dam, where the spillway meets with the rocky hillside. But the stilling basin is not nearly as naturalistic and seems awkward compared to the New Croton.

TITICUS RESERVOIR

The Titicus Dam was constructed between 1890 and 1896, and holds 7.2 billion gallons of water. No towns were submerged for its site, as large country estates covered most of the land. From here water is discharged to the Titicus River so that it can flow to the Muscoot Reservoir and then to the New Croton Reservoir. The dam was rebuilt in the 1990s.

PLATE 14

PLATE 15

PLATE 16

PLATE 17

PLATE 18

PLATE 19

PLATE 20

Croton System

PLATE 14 Spillway, East Branch Reservoir, Putnam County, New York, 1999
PLATE 15 Spillway, Croton Falls Diverting Reservoir, Putnam County, New York, 1994
PLATE 16 Croton Falls Diverting Reservoir, Putnam County, New York, 1999
PLATE 17 Connecting Channel Gatehouse, Putnam County, New York, 1995
PLATE 18 Connecting Channel Gatehouse, Putnam County, New York, 2000
PLATE 19 Dam and Spillway, Croton Falls Reservoir, Putnam County, New York, 1995
PLATE 20 Spillway, Croton Falls Reservoir, Putnam County, New York, 2000

Within the Croton system, several reservoirs were designed to work together to take advantage of different watersheds. The East Branch, Bog Brook, Croton Falls Diverting, and Croton Falls reservoirs are all connected by a tunnel and a channel cut when the reservoirs were built.

EAST BRANCH AND BOG BROOK RESERVOIRS

The East Branch and Bog Brook reservoirs were built as "Double Reservoir I." The watershed around the East Branch Reservoir was much larger than the one that served the Bog Brook Reservoir, so a ten-foot pipe was installed to connect the two. This allowed the Bog Brook Reservoir to fill with water from the East Branch's watershed. There were several small gatehouses and dams in this little subsystem, but the most interesting picture for me was where spillways from parts of the two reservoirs met. This view was taken from a highway bridge built many decades after the dams were completed.

CROTON FALLS DIVERTING RESERVOIR

The approach to the Croton Falls Diverting Reservoir takes you up alongside it; the long, stepped spillway that faces the road makes it clear that this is not a natural body of water. Then you see what looks like a mattress floating out in the middle, until you realize it is a concrete outlet for overflow. The reservoir is situated south of the Bog Brook and East Branch reservoirs, and is connected to the Croton Falls Reservoir by a channel. In the channel is a small gatehouse that controlled the flow from one reservoir to the other. It is no longer in use, but still in good condition. This is another of the sites in the system that I kept going back to—completely hidden from view, but not hard to get to once I knew where it was.

CROTON FALLS RESERVOIR

Croton Falls was the last reservoir to be built in the Croton system. Before work on it started in 1906, the town of Katonah had to be moved out of the way. By the time it was completed in 1911, the city's Board of Water Supply had begun work on a much larger system in the Catskills. Croton Falls as well as Cross River reservoirs were both designed and built by the Aqueduct Commission, and operated by the city's Department of Water Supply, Gas and Electricity.

I spent some time figuring out how to get to the top of the Croton Falls Dam, and when I did, I was rewarded by the view of a large expanse of land at the base. After I drove across the top, the road took me around and down, where I could see a very long spillway cut out of the side of the valley. Several years passed before I was able to get to the bottom of the spillway. Unlike many of the dams, a large section of land at the base of the dam is fenced off and I had to wait for permission for access to it. It seemed that the whole area around the reservoir had been designed to be a public space.

PLATE 21

PLATE 22

PLATE 23

PLATE 24

PLATE 25

PLATE 26

Croton System

PLATE 21 New Croton Dam, Westchester County, New York, 1999
PLATE 22 New Croton Dam, Westchester County, New York, 1999
PLATE 23 Spillway, New Croton Dam, Westchester County, New York, 1999
PLATE 24 Stilling Basin, New Croton Dam, Westchester County, New York, 1999
PLATE 25 Upper Gatehouse, New Croton Dam, Westchester County, New York, 2000
PLATE 26 Lower Gatehouse, New Croton Dam, Westchester County, New York, 2000

NEW CROTON DAM

All of the Croton system's eleven reservoirs and three controlled lakes feed into the New Croton Reservoir, which provides approximately ten percent of the city's water in normal times, and more in times of drought. The New Croton Dam took thirteen years to build, and was made largely by poorly paid immigrants, who lived in towns set up for them. Approximately 2000 residents had to be moved, and 1800 bodies in six cemeteries were dug up and reinterred. After the first five years of work much of the masonry construction of the dam had to be redone because of the poor quality of the work. This gave the Aqueduct Commission an opportunity to enlarge the dam by 290 feet in length. The reservoir is nineteen miles long, and five highway bridges had to be built to cross it.

I visited this dam more than any other site in the system. The base of it is located in a Westchester County park, and there is a road on the top of the dam, which made it very accessible. When the reservoir is overflowing, the spillway becomes a noisy waterfall, and the Croton River rushes by, much like it used to before the dam was built. When the only water coming down is through a few small leaks, the river looks more like a small pond. The spillway of the dam is sensitively integrated into the surrounding area; the way it meets the rocky hillside next to it seems almost natural.

After photographing the exterior of the dam for several years, I was finally permitted to go inside in 2000. From a door on top of the dam, I entered into what looked like an Egyptian tomb made of solid blocks of rock. It was a pigeon colony; there were nests, eggs, babies, and parent birds everywhere. Also inside was a room filled with gate valves and beautiful stone arches. Since the important controls for the dam were in another area, these rooms were not often visited by humans. The layout of the space seemed difficult to convey in a photograph, so I decided to work on one section rather than portray its entirety.

A smaller room at the bottom of the dam held a few additional gate valves. On entering, I saw an interesting but poorly lit space. Within a few minutes, the sun came out and it was as if someone had turned on a light switch. Because the ceiling was made up of metal grates with small rounds of glass, the light was focused on the wall and floor, creating what appeared to be some kind of coded message.

PLATE 27

PLATE 28

PLATE 29

PLATE 30

PLATE 31

Croton System

PLATE 27 New Croton Gatehouse, Westchester County, New York, 2000
PLATE 28 New Croton Gatehouse, Westchester County, New York, 2000
PLATE 29 New Croton Aqueduct Gatehouse, South Yonkers, New York, 1999
PLATE 30 Jerome Park Reservoir, Bronx, New York, 2000
PLATE 31 135th Street Gatehouse, New York, New York, 2001

NEW CROTON GATEHOUSE

Located on the New Croton Reservoir just a few yards from the original Old Croton Gatehouse (which is now preserved and submerged underwater), the New Croton Gatehouse is the third at the site, replacing also the gatehouse built at the beginning of the twentieth century. This building, completed in 1993, looks more like a part of the city's new water tunnel than a part of the city's oldest water system. Instead of being cramped, dark, and poorly lit, it is bright and clean, with poured concrete walls, stainless steel machinery, and pipes covered with colorful sound insulation. Here water from different sources is mixed so as to achieve the best taste.

SOUTH YONKERS GATEHOUSE

I had seen the beautiful drawings of the gatehouses on the New Croton Aqueduct. They showed all the underground works in great detail, but the only visible part of the South Yonkers Gatehouse is this small structure aboveground. I found it after searching for hours, and finally located it in an overgrown area behind some suburban houses.

JEROME PARK RESERVOIR

Jerome Park Reservoir was planned to be much larger; land originally designated for it is now filled by a subway train yard, parts of City University of New York's Herbert Lehman College, and several city high schools. As a balancing reservoir for the Croton system, it collects quite a bit of silt and other urban detritus, and has to be periodically emptied and cleaned.

135TH STREET GATEHOUSE

The New Croton Aqueduct was built between 1885 and 1891. The Terminal Gatehouse of the aqueduct still stands at 135th Street, though it is no longer connected to the system. Eight water mains, forty-eight inches in diameter, connected to the aqueduct at the gatehouse and carried water to the Central Park Reservoir. Water flow to the pipes was controlled in three separate ways to allow for maintenance work on any part at any time. The mains were buried deep enough to allow the Jerome Park Reservoir to be drained completely through the system. A connection was also made to the Old Croton Aqueduct at the gatehouse.

PLATE 32

PLATE 33

PLATE 34

Long Island System

PLATE 32 Gatehouse, Massapequa, New York, 1998
PLATE 33 Millburn Pumping Station, Freeport, New York, 1998
PLATE 34 Gatehouse, Wantagh, New York, 1997

The Long Island system was built between 1856 and 1858 to supply Brooklyn and Queens with water from wells and small lakes on Long Island. Small improvements were made through the late nineteenth century, and major improvements came in the 1890s. The main reservoir, on the Brooklyn/Queens border, was the Ridgewood Reservoir. It is still there but it has been drained and is now city parkland. There are quite a few remnants of the system, but most of them are either neglected artifacts or in ruins. A small gatehouse in Massapequa stands right next to the Long Island Railroad track, while one in Wantagh sits by an expressway, fenced in and nearly forgotten behind the weeds. There are also remains of the Millburn Pumping Station, once a magnificent romanesque building. It has not survived neglect and offered little to photograph except for one small building. The system went out of general use in the early 1960s.

PLATE 35

PLATE 36

PLATE 37

PLATE 38

PLATE 39

PLATE 40

PLATE 41

PLATE 42

PLATE 43

Catskill System

ASHOKAN RESERVOIR

The Catskill system was designed to include two collecting reservoirs—the Ashokan in Ulster County and the Schoharie in Schoharie, Delaware, and Greene counties. The two reservoirs are connected by the Shandaken Tunnel (the longest continuous tunnel in the world when it was completed in 1924) and Esopus Creek. From the Ashokan, the water travels ninety-two miles through the Catskill Aqueduct, to the Kensico Reservoir in Westchester County, and then through the Hillview Reservoir on its way to New York City. Another reservoir, Silver Lake in Staten Island, was the fifth reservoir in the system, but has been replaced by the largest underground storage tanks in the world.

Work on the Ashokan Reservoir took eight years and was completed in 1915 under the direction of Chief Engineer J. Waldo Smith. It was a huge project including the construction of earthen dikes, the Olive Bridge Dam, upper and lower gate chambers, the waste weir, and, of course, an aqueduct to bring the water to New York. Because of a drought in 1915 and periodic releases of water to test the aqueduct, it took some time for the reservoir to fill to its 123-billion-gallon capacity; many joked that the "Esopus Folly" would never hold water, and local newspapers questioned the need for it. Two thousand people in eight villages had to be moved to make way for the reservoir, and legal battles continued for years to determine what they would be paid for the loss of their homes and businesses.

OLIVE BRIDGE DAM

For years I tried to photograph the Olive Bridge Dam. I knew exactly where it was, I could drive over it, and I could see where it was from the reservoir side. But I could never see what it really looked like, since it was in the forest, far from the high fences and locked gates, and I did not have permission to go to it. Finally, in 1999, I was taken in to see it (by Water Supply Police on a break, no less) and had only a few minutes to try to make pictures.

One year later, I was permitted to photograph the dam again, this time accompanied by the reservoir's keeper. Though the excitement of the first visit was not there, I had much more time to photograph and was able to walk around the base of the dam and photograph some of the landscape.

UPPER GATE CHAMBER

The upper gate chamber is where the water begins its twenty-four-hour journey to Kensico Reservoir. As with many other structures in the system, there is much more below ground than above. The water enters through huge gates, then travels to the lower gate chamber for further screening, then into the Catskill Aqueduct. Large doorlike gate valves control the flow of water through the chamber. Stop shutters are used to close off the area around each valve for maintenance work. I photographed the empty shutter racks on one of the lower levels of the chamber.

DIVIDING WEIR

A weir divides the Ashokan Reservoir into two sections, at slightly different elevations, to take advantage of different size watersheds to the east and west of it. A roadway on top allows vehicles to cross over the reservoir.

PLATE 44

PLATE 45

PLATE 46

PLATE 47

PLATE 48

PLATE 49

Delaware System

PLATE 44 Spillway, Neversink Reservoir, Sullivan County, New York, 1999
PLATE 45 Spillway, Neversink Reservoir, Sullivan County, New York, 2000
PLATE 46 Spillway, Neversink Reservoir, Sullivan County, New York, 1999
PLATE 47 Spillway, Neversink Reservoir, Sullivan County, New York, 2000
PLATE 48 Stilling Basin, Neversink Reservoir, Sullivan County, New York, 1999
PLATE 49 Spillway, Merriman Dam, Rondout Reservoir, Ulster County, New York, 1999

NEVERSINK RESERVOIR

Construction on the Neversink Reservoir in Sullivan County began in 1941 but was stopped in 1943 because of World War II. Work did not resume until 1946 and was completed in 1953. The Neversink connects to the Rondout Reservoir via the six-mile Neversink Tunnel. The first time I visited it I photographed the spillway from the highway bridge that crosses over it. The bridge was so shaky that I had to time my exposures so that no vehicles shook my tripod. Looking down at the spillway it was easy to underestimate how large it was. When I finally had a chance to get close, what looked like a small step from above was over ten feet high.

RONDOUT RESERVOIR

The Rondout Reservoir, in Ulster and Sullivan counties, is one of New York's most important reservoirs; water from the Pepacton, Cannonsville, and Neversink reservoirs flows through here before traveling to the Kensico Reservoir, eighty-five miles away via the Delaware Aqueduct. The Rondout was built between 1937 and 1954, with a three-year break during World War II. As with most of the other modern reservoirs, it was created by the construction of earthen dams. Aside from the gatehouse, the spillway is the only stone construction visible to the passerby.

PLATE 50

PLATE 51

PLATE 52

PLATE 53

Delaware System

PLATE 50 Spillway, Downsville Dam, Pepacton Reservoir, Delaware County, New York, 1997

PLATE 51 Spillway, Downsville Dam, Pepacton Reservoir, Delaware County, New York, 1997

PLATE 52 Spillway Gatehouse, Downsville Dam, Pepacton Reservoir, Delaware County, New York, 1997

PLATE 53 Spillway, Stilesville Dam, Cannonsville Reservoir, Delaware County, New York, 1997

CANNONSVILLE AND PEPACTON RESERVOIRS

These are the city's two newest reservoirs and continue the trend toward less visible masonry and more earth-covered structures. Together, the two reservoirs hold 236 billion gallons of water; the Pepacton is the city's largest. Both were planned before World War II, but their construction did not start until the war had ended. Construction was also delayed by a U.S. Supreme Court battle to determine how the Delaware watershed's supply would be split among New York, New Jersey, and Pennsylvania; a tri-state commission now regulates water use. The reservoirs are in areas that are still relatively undeveloped, though the water quality here is threatened by an expanding population upstate. New York is under federal mandate to protect a large area within and around these watersheds if it is to avoid filtration.

PLATE 54

PLATE 55

PLATE 56

PLATE 57

PLATE 58

Delaware System

PLATE 54 Kensico Dam, Valhalla, New York, 1999
PLATE 55 Kensico Dam, Valhalla, New York, 1999
PLATE 56 Kensico Dam, Valhalla, New York, 1999
PLATE 57 Aerator Field, Kensico Reservoir, Valhalla, New York, 1994
PLATE 58 Delaware Aqueduct Influent, Kensico Reservoir, Valhalla, New York, 2000

KENSICO RESERVOIR

Kensico Reservoir was built to receive water from the Catskill system's Ashokan and Gilboa reservoirs, but now can receive Delaware system water as well. It was completed in 1915 and engulfed the smaller Kensico Reservoir that was also part of the Williamsbridge water supply. Kensico is the place where chlorination and other treatment occurs, though the water is not filtered here or any place else in the system. (New York must build a filtration plant for Croton water, but so far has managed to avoid filtering Catskill and Delaware water.) The reservoir can hold over thirty billion gallons of water, about a twenty-six-day supply for the city. While not a collecting reservoir, it does receive runoff from neighboring communities and is endangered by road expansion and other development. Water travels from here to the Hillview Reservoir, which is used to balance water supply on an hourly basis.

Two aerator fields were built as part of the Catskill system, one at Ashokan and the other at Kensico, both used to help purify the water. The aerator at Ashokan has been replaced by a pool and sculpture, but Kensico's, though long out of use, remains.

PLATE 59

PLATE 60

PLATE 61

PLATE 62

PLATE 63

PLATE 64

PLATE 65

PLATE 66

PLATE 67

Dewatering and Other Miscellaneous Waterworks

PLATE 59 Dewatering Apparatus, Shaft 6, Delaware Aqueduct, Chelsea, New York, 2000
PLATE 60 Shaft 6, Delaware Aqueduct, Chelsea, New York, 2000
PLATE 61 Shaft 6, Delaware Aqueduct, Chelsea, New York, 2000
PLATE 62 Shaft 9, Delaware Aqueduct, West Branch Reservoir, Putnam County, New York, 1997
PLATE 63 Roof, Dewatering Shaft, Catskill Aqueduct, Westchester County, New York, 2000
PLATE 64 Gatehouse, Hillview Reservoir, Yonkers, New York, 2000
PLATE 65 Bypass Tunnel, Hillview Reservoir, Yonkers, New York, 2000
PLATE 66 Dewatering Apparatus, Shaft 21, City Tunnel No. 1, New York, New York, 2000
PLATE 67 Dewatering Apparatus, Shaft 21, City Tunnel No. 1, New York, New York, 2000

DEWATERING APPARATUS, SHAFT 6, DELAWARE AQUEDUCT

This dewatering shaft was completed in the 1940s. It serves the eighty-four-mile Delaware Aqueduct that was completed in 1945 and that travels under the Hudson River at a depth of about seven hundred feet. From Shaft 6 the city will use an unmanned submarine to investigate a decades-old leak in the Delaware Tunnel.

SHAFT 9, DELAWARE AQUEDUCT, WEST BRANCH RESERVOIR

When the Delaware Aqueduct was built, its route provided an opportunity to connect to the West Branch Reservoir and tie in to the Croton system. Water can flow into the New Croton Reservoir, or bypass the Croton system and drain into Kensico Reservoir.

GATEHOUSE, HILLVIEW RESERVOIR

I once asked E. L. Doctorow if he had used any part of the water system as a model for the structures in his novel *The Waterworks*. He told me that he had the gatehouse at Hillview Reservoir in mind, though he was not sure that it was actually a water supply building. He was right about the building, but it did not exist at the time his story takes place.

BYPASS TUNNEL, HILLVIEW RESERVOIR

The Hillview Reservoir in Yonkers is the last stop for much of New York's water before it reaches the city. As with so many other components of the water system, there are bypasses and redundancies built into it. The Bypass Tunnel that runs through it allows water to skip the reservoir and go directly into the city tunnels. When this photograph was taken, work was progressing to strengthen the tunnel. The circles of light at the bottom of the tunnel are created by holes made in the top of it during renovation.

DEWATERING APPARATUS, SHAFT 21, CITY TUNNEL NO. 1

More than seven hundred feet below ground, City Tunnel No. 1 crosses under the East River to Brooklyn. In the event that repairs on the tunnel were needed, workers could descend to it in a capsule known as a dewatering apparatus. A crane would lower the capsule into the shaft, and pumps would begin to empty the water. As the water was removed, the capsule would descend, floating down the shaft. The water would be pumped out through pipes above the capsule; additional sections of pipe would be added as the capsule went lower and lower, until it finally reached the bottom. However, in all the years of the dewatering facility's existence, it had never been used.

I waited five years to visit this site. I saw it across the East River from my Brooklyn studio almost daily, and I saw it every time I drove onto the FDR Drive to go to the reservoirs up north. I had seen pictures of its interior. I was told several times that it was about to be demolished, and that there was probably nothing left. In fact, when I did get in, my timing was perfect. It was going to be demolished, but much of the junk inside had been removed, so I could see the structure in the dewatering shaft house. Now that it has been demolished, portable machines are available for the same function.

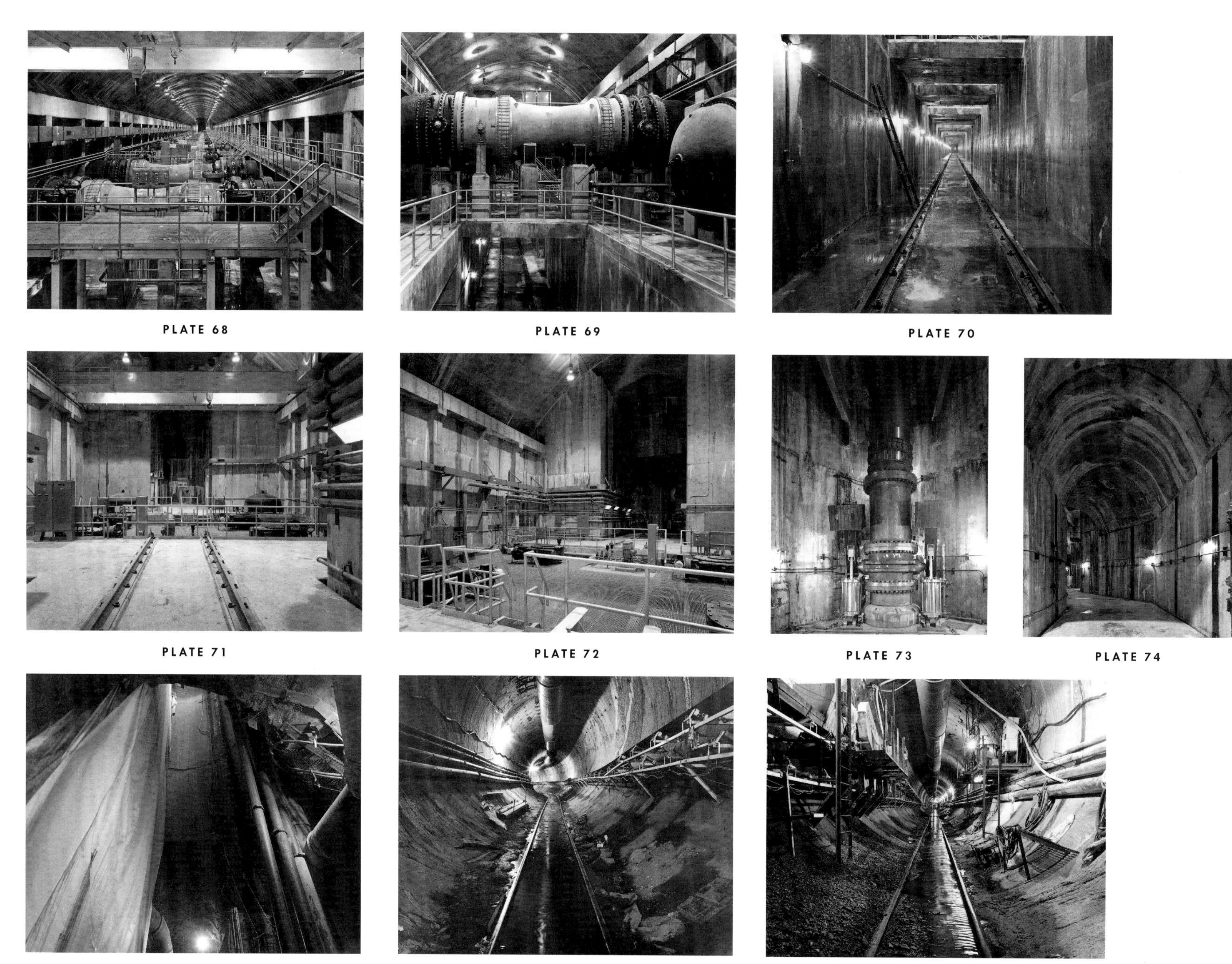

PLATE 68

PLATE 69

PLATE 70

PLATE 71

PLATE 72

PLATE 73

PLATE 74

PLATE 75

PLATE 76

PLATE 77

PLATE 78

PLATE 79

PLATE 80

City Tunnel No. 3

PLATE 68 Shaft 2b, City Tunnel No. 3, Bronx, New York, 1992
PLATE 69 Shaft 2b, City Tunnel No. 3, Bronx, New York, 1992
PLATE 70 Shaft 2b, City Tunnel No. 3, Bronx, New York, 1992
PLATE 71 Shaft 13b, City Tunnel No. 3, New York, New York, 1994
PLATE 72 Shaft 13b, City Tunnel No. 3, New York, New York, 1994
PLATE 73 Shaft 15b, City Tunnel No. 3, Roosevelt Island, New York, New York, 2000
PLATE 74 Shaft 15b, City Tunnel No. 3, Roosevelt Island, New York, New York, 2000
PLATE 75 City Tunnel No. 3, Queens, New York, 1997
PLATE 76 City Tunnel No. 3, Queens, New York, 1998
PLATE 77 City Tunnel No. 3, Queens, New York, 1998
PLATE 78 City Tunnel No. 3, Queens, New York, 1998
PLATE 79 City Tunnel No. 3, Queens, New York, 1997
PLATE 80 City Tunnel No. 3, Brooklyn, New York, 1998

New York City began construction of its third water tunnel in 1970. The tunnel is one of the largest municipal construction projects ever undertaken and will take at least fifty years to complete. It is being built so that the two older distribution tunnels completed in 1915 and 1933, can be temporarily shut down for inspection and maintenance work, which has not been done since their construction. The tunnel is now cut by a giant boring machine (the "mole") instead of by explosives, as it was in the past. Once the rocks and other debris are removed, a steel form is put in place and the concrete is poured, creating a tunnel that may be as large as twenty feet in diameter. After the concrete sets, the form is folded up and moved to the next section.

I had the opportunity to photograph several parts of City Tunnel No. 3 under construction. To visit them, I had to wear a rain suit, steel-tipped boots, hardhat, ear and eye protection, and an oxygen cartridge. I then descended in the elevator (the "cage") about eight hundred feet, the light gradually disappearing until I reached the bottom. There I put my equipment on my back and waited for a small supply train to take me a few miles into the tunnel. Once I walked two miles before the train came. Long exposures were often necessary, not always easy when the train was coming through or when a crew of "sandhogs" had to get by. The completed valve chamber in the Bronx is an expanse of concrete and stainless steel, a control center for City Tunnel No. 3. It is entered through a thick stainless steel door, which was designed during the cold war to withstand anything meant to destroy it. Down an elevator about 250 feet is an immaculate concrete space filled with steel valves that determine where the water will go. The temperature is a constant fifty-eight degrees Fahrenheit. Under Central Park and Roosevelt Island there are smaller valve chambers that allow sections of the tunnel to be shut down for maintenance; more will be built later as the tunnel reaches other parts of the city.

SUGGESTED READINGS

Aqueduct Commission. *Report to the Aqueduct Commissioners, 1883–1887*. New York: Aqueduct Commission, City of New York, 1887.

__________. *Report to the Aqueduct Commissioners, 1888–1894*. New York: Aqueduct Commission, City of New York, 1894.

__________. *Report to the Aqueduct Commissioners, 1895–1907*. New York: Aqueduct Commission, City of New York, 1907.

Blake, Nelson Manfred. *Water for the Cities: A History of the Urban Water Supply Problem in the United States*. Syracuse, N.Y.: Syracuse University Press, 1956.

Board of Water Commissioners, City of Brooklyn. *The Brooklyn Water Works and Sewers: A Descriptive Memoir*. New York: D. Van Nostrand, 1867.

Board of Water Supply in the City of New York. *Long Island Sources, Volumes 1 and 2*. New York: City of New York, 1912.

Burr, William H., Rudolph Hering, and John R. Freeman. *Reports in Relation to Providing an Additional Water Supply for the City of New York*. New York: New York City Commission on Additional Water Supply and M. B. Brown Company, 1905.

Burrows, Edwin G., and Mike Wallace. *Gotham: A History of New York City to 1898*. New York: Oxford University Press, 1999.

Kunz, George Frederick. *Catskill Aqueduct Celebration*. New York: City of New York, 1917.

Dawson, Robert, Gerald Haslam, and Stephen Johnson. *The Great Central Valley: California's Heartland*. Berkeley: University of California Press, 1993.

Department of City Works, City of Brooklyn. *Report on Future Extension of Water Supply for the City of Brooklyn*. Brooklyn: City of Brooklyn, 1896.

Doctorow, E. L. *The Waterworks*. New York: Random House, 1993.

Department of Water Supply, Gas and Electricity. *The Water Supply of the City of New York*. New York: City of New York, 1939.

__________. *The Water Supply of the City of New York*. New York: City of New York, 1950.

Evers, Alf. "The Ashokan Reservoir is Built." 1983. In *A Hudson Valley Reader*, edited by Bonnie Marranca. Woodstock, N.Y.: Overlook Press, 1995.

FitzSimons, Neal, editor. *The Reminiscences of John P. Jervis, Engineer of the Old Croton*. Syracuse, N.Y.: Syracuse University Press, 1971.

Frontinus, Sextus Julius. *The Water Supply of the City of Rome*. A.D. 97. Translation by Clemens Herschel. Boston: New England Water Works Association, 1973.

Galusha, Diane. *Liquid Assets: A History of New York City's Water System*. Fleischmanns, N.Y.: Purple Mountain Press, 1999.

Gandy, Matthew. *Concrete and Clay: Reworking Nature in New York City*. Cambridge, Mass.: MIT Press, 2002.

Glenn, C. Austin. *New York City Reservoirs in the Catskill Mountains, Guide to Fishing Waters*. Ithaca, N.Y.: Outdoor Publications, 1995.

Goldman, Joanne Abel. *Building New York's Sewers: Developing Mechanisms of Urban Management*. West Lafayette, IN: Purdue University Press, 1997.

Greenberg, Stanley. *Invisible New York: The Hidden Infrastructure of the City*. Baltimore: Johns Hopkins University Press, 1998.

Hall, Edward Hagaman. *The Catskill Aqueduct and Earlier Water Supplies of the City of New York*. New York: Mayor's Catskill Aqueduct Celebration Committee, 1917.

Hodge, A. Trevor. *Roman Aqueducts and Water Supply*. London: Duckworth, 1992.

Hudson River Museum. *The Old Croton Aqueduct: Rural Resources Meet Urban Needs*. Yonkers, N.Y.: Hudson River Museum, 1992.

Jackson, Donald C. *Great American Bridges and Dams*. Washington, D.C.: Preservation Press, 1988.

Kazin, Alfred. *A Walker in the City*. New York: Grove Press, 1951.

Koeppel, Gerard T. *Water for Gotham: A History*. Princeton: Princeton University Press, 2000.

Larkin, Daniel F. *John B. Jervis, An American Engineering Pioneer*. Ames, Iowa: Iowa State University Press, 1990.

Ondaatje, Michael. *In the Skin of a Lion: A Novel*. Toronto: McClelland and Stewart, 1987.

Sive, Mary Robinson. *Lost Villages: Historic Driving Tours in the Catskills*. Delhi, N.Y.: Delaware County Historical Association, 1998.

Steuding, Bob. *The Last of the Handmade Dams*. Fleischmanns, N.Y.: Purple Mountain Press, 1989.

Tower, F. B. *Illustrations of the Croton Aqueduct*. New York: Wiley and Putnam, 1843.

Wegmann, Edward. *The Water Supply of the City of New York, 1658–1895*. New York: John Wiley and Sons, 1896.

__________. *Conveyance and Distribution of Water for Water Supply*. New York: D. Van Nostrand, 1918.

__________. *The Design and Construction of Dams, Eighth Edition*. New York: John Wiley and Sons, 1927.

Weidner, Charles C. *Water for a City*. New Brunswick, N.J.: Rutgers University Press, 1974.